From Gutenberg to Google

From Gutenberg to Google

The History of Our Future

TOM WHEELER

BROOKINGS INSTITUTION PRESS
Washington, D.C.

1775 Massachusetts Avenue, N.W., Washington, D.C. 20036
www.brookings.edu

The Brookings Institution is a private nonprofit organization devoted to research, education, and publication on important issues of domestic and foreign policy. Its principal purpose is to bring the highest quality independent research and analysis to bear on current and emerging policy problems. Interpretations or conclusions in Brookings publications should be understood to be solely those of the authors.

Library of Congress Cataloging-in-Publication data are available.
ISBN 978-0-8157-3532-8 (cloth : alk. paper)
ISBN 978-0-8157-3533-5 (ebook)

9 8 7 6 5 4 3 2 1

Typeset in Bulmer MT

Composition by Elliott Beard

For Melvin, Hunter, and Skyler—
it is you who will decide the future

Contents

Acknowledgments

Typically, after thanking those who helped in the creation of a book, the author adds a final note of thanks to his wife. For me, it is the place to start. Absent the love and support of Carol Wheeler, my life would be resoundingly less rich, and I most certainly would be a different person. Her wisdom is a North Star. Her love has shaped my existence. Within that broad scope, this book is a trifle, yet producing it required a shared commitment. Patience, understanding, and good humor are required as your husband retreats into research or pecks away at his computer, tasks go undone, and you are ignored. Tolerance is tested when an obscure discovery captures his imagination and dinner table conversations continually revert to that fascination. But beyond all that, Carol is the editor in chief: it doesn't get in the book until it makes sense to her. Thank you to a great lady who plays so many roles so well!

While I was at the Federal Communications Commission, dealing with changes imposed by technology, the echoes of history were everywhere. The last sections of the book were stimulated and influenced by that experience. It was an experience I was fortunate to share with an amazing team who together wrestled with our role in building the future. Every morning I would gather with Ruth Milkman, Phil Verveer, Jon Sallet, and Louisa Terrell to tackle the challenges. We were fortunate to be able to call on Gigi Sohn, Diane Cornell, and Howard Symons and to be able to turn to bureau and office chiefs Dave Simpson, Roger Sherman, Jon Wilkins, Julie Veach, Matt DelNero, Travis LeBlanc, Alison Kutler, Shannon Gilson, Julius Knapp, Bill Lake, and Mindel De La Torre, along with Gary Epstein, Kim Hart, Sagar Doshi, and Emmaka Porchea-Veneszee. Probably the toughest job in the chairman's office is the legal adviser's; Maria Kirby, Daniel Alvarez, Renee Gregory, Jessica Almond, Stephanie Weiner, "Smitty" Smith, and Holly Sauer all filled that role with distinction. It was a privilege to work with and learn from these talented individuals, all of whom helped shape this book by shaping how we dealt with the challenges the new network technologies were imposing on consumers and the market.

A writer is very fortunate when he has a substantive sounding board. Jon Sallet, legal intellect and eloquent writer, was that gift for me. As general counsel of the FCC, Jon was intimately involved in how we applied the lessons of history to shaping the future. After leaving the FCC, Jon was the sounding board for just how these experiences could be reflected in the book.

Blair Levin is a font of stimulating insights and the one from whom I lifted the wonderful alliteration that the internet is the "most powerful and pervasive platform in the history of the planet." Bob Barnett has been indefatigable in his support of this and other undertakings. Marion Maneker, editor of an earlier book of mine, helped me sharpen the original idea. Tom Standage, whose wonderful books link history and current events, was an inspiration. Scott Jordan, former CTO of the FCC, kept me on the straight and narrow technically (any errors are mine, not his). Dipayan Ghosh provided great counsel and review. Tom Schwartz, Joel Swerdlow, Robert Roche, Rajeev Chand, Rick Stamberger, Kathy Brown, Susan Crawford, Kevin Werbach, and Lawrence Yanovitch contributed insights, facts, and counsel. Matthew Spector came into the process late as research assistant and, among other contributions, broke me of the habit of capitalizing the word "internet." Marjorie Pannell's eagle eye caught several infelicities in my prose. Heidi Fritschel's performed the same service during proofreading.

In my post-FCC life, I have been fortunate to be able to turn to Darrell West of the Brookings Institution and Nicco Mele of the Shorenstein Center at the Harvard Kennedy School. I am grateful to them for providing me the opportunity to work on this and other projects.

And to Bill Finan of Brookings Institution Press, thank you for your faith in this book.

Preface

When President Obama asked me to be chairman of the Federal Communications Commission, I was in the midst of research on a book tracking the creation and impact of new technology-driven networks. Taking over the federal agency responsible for the nation's networks (approximately one-sixth of the national economy), I moved from a career in new technology and an observer of technology-driven change to suddenly becoming a participant in shaping society's response to such change. As the economy evolved from old analog practices to a new digital reality, I found myself dealing with twenty-first-century iterations of the same kinds of issues I had been studying.

The opportunity to put into practice the lessons of historical precedent led to a series of decisions that themselves were historic. The Open Internet Rule (sometimes called net neutrality) dealt with the well-documented incentive of networks to discriminate against users for commercial gain. The kinds of privacy responsibilities that applied to traditional telephone networks were extended to internet access networks. Because networks have always been routes of attack, a new regulatory process was developed to coordinate and oversee the cybersecurity hygiene practices of digital networks, and cyber expectations were established for the next-generation wireless network.

Unfortunately, these and other initiatives have been repealed by the Trump FCC or the Republican Congress. While such a result is woeful, it fits with the narrative of this book that network-driven change is a bumpy road. What makes a nation great is not a retreat into hazy recollections of the "good old days" but rather how it responds to the challenges of changed circumstances, including the impact of new technology. Historically, such responses have occurred in fits and starts. Ultimately, new rules and practices embrace the reality of the future over the practices of the past—but getting there isn't easy.

The networks that connect us are a force that defines us. The telephone, internet, and broadcasting, satellite, and cable networks regulated by the FCC may constitute one-sixth of the American economy, but they are relied on by the other five-sixths to do business and by every individual in his or her daily life.

This book looks at the pattern of such network-driven outcomes over time. The first sections tell the stories of how the great network technologies came to be. They also recount society's response to the changes—those who rose in opposition and those who saw opportunity. Finally, each of these chapters draws a Darwinian connection between the earlier technology and today's technology.

While new networks may be the primary enabling force, history shows it is the secondary effects of such networks that are transformational. The last section of the book, therefore, looks at what is happening today and what is developing for tomorrow. The penultimate chapter reports on a select group of ongoing effects of today's technology. While it is a subjective selection of topics based on personal experiences, it is nonetheless representative of our ongoing challenges.

The final chapter looks at four network-driven forces that will transform tomorrow. The combination of low-cost computing and ubiquitous networks has changed the primary activity of networks from transporting information to orchestrating information to creating something new. Combining ubiquitous computing and Big Data has created artificial intelligence. The distributed network has replaced the traditional paradigm of centralized trust with the distributed trust of blockchain. And hanging over everything is the challenge of cybersecurity.

The privilege of being FCC chairman gave me the opportunity to try to relate the lessons of the past to some of the most high-profile and important network-related disputes of today. The book you hold reflects that experience. I hope to show you the power of network revolutions, the fears they foster, and the opportunities they create. I want to illustrate how some of my own decisions in office were affected by the lessons of the past. Most of all, I want to convince you that the most important impact of network revolutions is not the network technology but how society reacts to that technology—and how that is something we control.

Tom Wheeler
June 2018

Prologue

"Move Fast and Break Things." The message was ubiquitous as I walked through Facebook's offices. Neatly printed signs proclaimed the admonition, as did freehand felt-pen scrawlings or cut-out letters. The gospel was everywhere: in hallways, stairwells, break areas, and workspaces.[1]

Indeed, Facebook and its internet cohorts have broken things at an amazing pace. Fifty-two percent of companies in the Fortune 500 at the turn of the twenty-first century don't exist anymore.[2]

The largest taxi company owns no vehicles.

The largest accommodations firm owns no hotels.

Associated Press stories on baseball games and corporate earnings are composed without human involvement as computer programs turn statistics into words to create journalism.[3]

Teenagers' applications for driver's licenses are down. Why bother? Online constant connectivity and on-demand transportation provide independence without the parallel parking test.[4]

Google is better informed about health outbreaks than the Centers for Disease Control and Prevention (CDC). As the infected go online to check their symptoms, Google's algorithms identify and track health trends long before doctors report to the CDC.

Inanimate objects are talking to us. An umbrella sends a text message you are about to leave it behind. A dog bowl signals it is time for Fido's walk by reporting his water consumption. A tampon signals it needs to be changed.[5]

And autonomous cars driving down the highway symbolize the heretofore unimaginable new realities that result when tens of billions of microchips embedded into everything flood the world with never-before-seen amounts of data, to be orchestrated by computer intelligence into completely new products and services.

Yes, we are moving fast and breaking things. We sit astride the most powerful and pervasive platform in the history of the planet[6]: the combination of low-cost, ever more powerful computing power and ubiquitous digital connectivity.

How did we get here? What does it mean?

We Have Been Here Before

Our new network technology may be the most powerful and pervasive in history, but it is not the first time new networks have confronted individuals and institutions with massive change. We should not delude ourselves into believing that somehow we are experiencing the greatest technology-driven changes in history—at least not yet.

We have been here before. What we are presently experiencing is history's third great network revolution.

The original information network was Johannes Gutenberg's fifteenth-century discovery of movable-type printing. The network of printers that sprang up across Europe ended the monopoly of information that priests and princes had exploited in pursuit of power. The free movement of ideas fired the Reformation, spread the Renaissance, and became the basis of all that followed.

Four centuries passed before the next great network-driven transformation appeared. This time it was a pair of symbiotic networks: the railroad and the telegraph. Steam locomotives vanquished the geographic distances that had always defined the human experience. As if that weren't revolution enough, the telegraph simultaneously eliminated time as a factor in the delivery of information. As one historian graphically described it, the resulting upheaval imposed the paradox of people living their lives "with one foot in manure and the other in the telegraph office."[7]

Viewed in context, the changes of the twenty-first century do not yet measure up to the effects of printing, steam power, and messages by sparks. Today's "revolutionary" technologies are a continuation of those earlier discoveries. While the new technologies have shown hints of transformative powers, we can only forecast an expectation of the true transformation that is coming.

The network technologies that are changing our today and defining our tomorrow are part of a Darwinian evolution. Technologically, each of the earlier network revolutions was a building block to the networked technologies of today. Sociologically, the angst and anger occasioned by today's upheavals track with the sentiments of earlier eras.

Reverse-engineer the TCP/IP language of the internet and you'll find Gutenberg's intellectual breakthrough for expressing information.

Track the history of the computer microchip and you'll end up in the era of steam and the world's first commercial railway. At a time when replacing muscle power with steam power was creating the Industrial Revolution, the idea of replacing brain power with machinery presaged the computer revolution.

Consider the off-on signals of the binary digital network and discover the dot-dash of the telegraph.

Amid these earlier technological changes, fear, resistance, and pushback were ever present. The railroad, for instance, was "an unnatural impetus to society," one journalist concluded, that would "destroy all the relations that exist between man and man, overthrow all mercantile regulation, and create, at the peril of life, all sorts of confusion and distress."[8]

These are the stories this book explores. We didn't reach today by accident, and that journey is important to appreciating what we're doing and where we're going.

The "Good Old Days" Weren't

The involuntary imposition of technology-driven change severs today from many of the anchors that previously provided stability and security. In reaction, a desire for the "good old days" manifests itself in everything from the ballot box to the nostalgic marketing of products.

The good old days, however, were far from idyllic—yet they produced greatness.

Throughout the stories of the earlier network revolutions, opposition was rampant as tradition was upset by economic insurgency and social insecurity. While attention tends to focus on the new technology itself, history makes it clear that it is the secondary effects of the primary technology that are transformative. And the transformation is inherently difficult because, by definition, neither the technology nor its effects are sufficiently mature to effectively substitute for the institutions they are disrupting. The history of new technology is the often painful process of reaching such maturity, including dealing with the opposition of those whose interests are threatened.

When Rupert Murdoch warned about the internet's threat to publishing,[9] for instance, he sounded very much like the sixteenth-century Vicar of Croydon warning "We must root out printing or printing will root out us."[10] Similarly, when today we complain about how constant connectivity is dominating our lives, we echo Henry David Thoreau's lament that "we do not ride on the railroad, it rides on us,"[11] or the warnings of nineteenth-century doctors who argued that by upsetting nature's natural rhythm, the "whirl of the railways and the pelting of telegrams" would produce mental illness.[12]

While the difficulties and struggles initiated by the earlier networks have been buffed smooth by the sands of time, we should not delude ourselves with idyllic images of golden bygone eras devoid of network-initiated pain, pathos, and struggle.

Relying on gauzy images of the past and our limited calendar of personal experience to make judgments about our own circumstances obscures the essential fact that we aren't alone in facing these challenges. Limiting our horizons by ignoring our history denies us an essential appreciation: that the greatness of a people comes not from a retreat into halcyon memory but from the advances they make as they respond to newly created challenges.

This book tells that history through the stories of the step-by-step creation of the technologies at the root of our new realities, as well as through the insight those stories provide into how earlier generations responded when confronted by destabilizing new technology. It is now our turn to craft stability out of technological tumult. The last section of the book addresses a sample of such modern challenges.

Parallel Paths to Today

The route to today's reality followed two parallel paths. Down one path progressed the almost 200-year stop-and-start development of computing power. In 1965 this history had a defining moment when Intel cofounder Gordon Moore forecasted that the capabilities of a new product called a microprocessor would double every eighteen to twenty-four months. "Moore's law" has defined the pace in the half century since.

As Moore's law formulates, the computer chip in your pocket or purse is a thousand times more powerful than the chip of only twenty years ago. The computing power that once required a multimillion-dollar supercomputer now lives on your phone. While Moore's law has begun to slow, its trajectory continues, with the result that the computer in your pocket tomorrow (or the chip in your toothbrush, shipping pallet, or light bulb) will be exponentially more powerful—and less expensive—than what we know today.

Over the same period, on a parallel communications path, the concept of electronic network connectivity progressed from messages delivered by telegraph sparks, to Alexander Graham Bell's replication of the human voice across a universal network, to the zeros and ones of the digital network.

When modems made computer digital code into sound, the phone network became the pathway for computer connectivity. In 1969 four research universities connected their computers through phone lines as part of a project funded by the U.S. Defense Department's Advanced Research Project Agency (ARPA). Dubbed ARPANET, it was the internet's opening act.

Then computing and communicating had sex.

A result of the combination of the two paths was the seeming disappearance of the technologies. For a century and a half, radio was separate from the wired phone network; then, as we will see, computers allowed a user to jump between low-power radio antennas. Bell's network leapt off the wires to vanish into the wireless ether. In a similar manner, computing moved from devices parked in special rooms or on desktops into fingernail-sized microprocessors, and ultimately disappeared into the cloud. The result—ever more powerful computing interacting over ubiquitous communications networks—has created the essential commodity of the twenty-first century.

Our Moment, Our Challenges

With this new communicating commodity have come wonderful and expansive new capabilities—as well as an equally expansive collection of challenges.

We can no longer escape. Once, being out of the office or away from home was an opportunity to bail out. Now you can be away but never apart.

The new reality of never being out of touch has boosted productivity and convenience, but at the price of personal freedom.

Jobs disappear. Industrial companies that once employed thousands yield to internet companies with only a handful of employees. In 2012 the venerable photography company Kodak, which had once employed 165,000 people, went bankrupt. That same year, the internet photo-sharing service Instagram, with fifteen employees, sold for $1.2 billion.[13]

Privacy expectations disappear. We leave digital tracks wherever we go and whatever we do. The new capital of the twenty-first century is such digital information. When so-called Big Data tracks diseases more quickly, or shares genomic data to advance science and industry, it moves society forward. The same technology, however, also invades our private space by sucking up personal information to be bought and sold for corporate profit.

Community is threatened. The founding fathers expressed their faith in a nation that is the sum of its parts with the national motto *E Pluribus Unum* (Out of Many, One). The networks that connect us today are having a "de-*unum*" effect by exploiting software algorithms to disassemble the shared information experiences that are necessary for a republic to succeed.

New market dominance is created. The new distributed network technology has pushed network applications away from centralized hubs while perversely creating a new kind of recentralization and market power. As the digital networks distribute activity, they collect and create information about network users. The aggregation of such information by a handful of companies thus becomes a new bottleneck to the operation of a free and competitive market.

Challenges such as these make up our historical moment. Just as we judge previous generations by how they handled their period of change, so shall we be judged.

Part I

Perspective

If history were taught in the form of stories, it would never be forgotten.

—Rudyard Kipling

One

Connections Have Consequences

The marriage of computing and communications was a shotgun wedding. This time, however, the shotgun was a nuclear bomb.

At the height of the Cold War, the United States relied on the telephone network to deliver commands to its nuclear strike forces. This meant, however, that the launch of bombers and missiles was vulnerable. Because the telephone network was a series of centralized hubs at which messages were switched from one path to another, all an adversary had to do was take out a few of those hubs and the nation's ability to launch a retaliatory attack would be impaired.

The U.S. government commissioned a California think tank, the RAND Corporation, to develop a solution for this soft spot. RAND's answer was a new network architecture that eliminated the vulnerable central switching points. The new network resembled a fishnet. If one knot on the fishnet was eliminated, there were multiple other routes the message could follow to reach its final destination.

The new network typology: centralized, decentralized, and distributed networks.

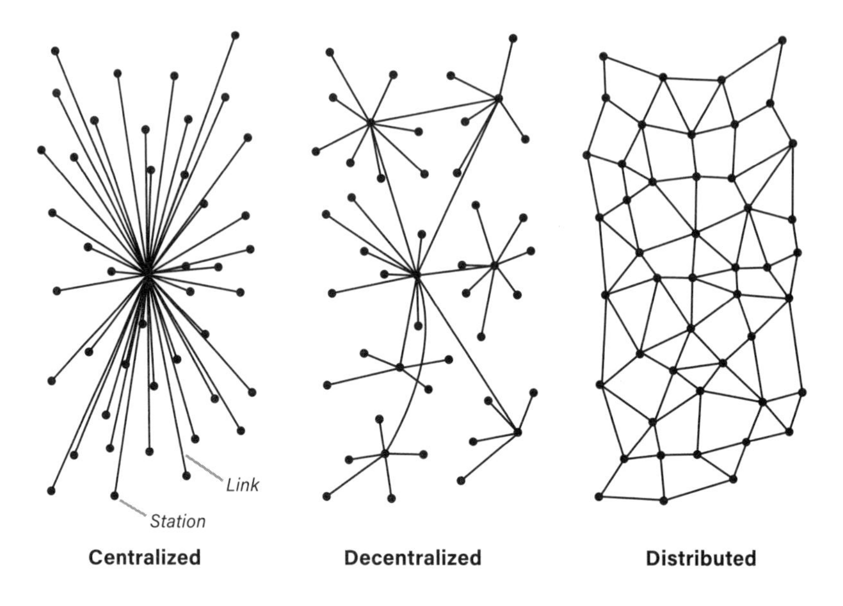

Source: Paul Baran, "On Distributed Communications: I. Introduction to Distributed Communications Networks," Memorandum RM-3420-PR (Santa Monica, Calif.: RAND Corporation, August 1964). Reproduced by permission.

The visionary behind this idea was a Polish immigrant named Paul Baran. In his 1964 paper, Baran proposed digitizing a voice phone call and then breaking that digital information into small packets of data. Instead of being sent over an end-to-end telephone circuit, the packets would be dispatched into a network of interconnected computers that would read the packet's address and then pass it on to the next computer in the direction of its destination. If one computer was knocked out, the packets would work their way around the problem by being re-sent to other nodes.[1]

It was an idea as big as the bomb itself.

Communicating computers handing packets of data to each other across a distributed network would become the hardware and software model driving our current network revolution. Contrary to urban legend, it was not the internet. However, digitizing information into packets in order to move the functions of the network out of central points and closer to the network's edge is the technological concept that underpins the internet.

Paul Baran's visualization of a new network architecture reconfigured network concepts that had existed for millennia. It started us on the path to the third great network revolution.

We are a network-centric species; the networks that connect us have always defined us. The most powerful external force in the human experience is the manner in which we link ourselves together. The most transformative technologies, therefore, have been those that changed the nature of that interconnection.[2]

The early networks were built around nature—rivers, mountains, even continents. Human social structures formed to exploit these natural networks, as well as to defend against network-based threats as diverse as starvation and war. As the basic technologies of life expanded, however, the flow of information remained limited by crude communication tools.

The first technology-based information network did not appear until the fifteenth century with the advent of the movable-type printing press.[3] Its arrival hastened the end of the medieval world and the birth of the modern era.

For centuries, knowledge had been kept barricaded within handwritten manuscripts. A stable and secure container for cultural and scientific information, these records required a large, expensive infrastructure to be produced and maintained. To accomplish this, nobles and priests, who made up "the Establishment," constructed a thick vault of high costs and mystic traditions around their priceless libraries. It was a system that not only protected the knowledge but also exploited it to perpetuate its owners' position.

Johannes Gutenberg picked the lock that had kept knowledge confined for centuries. The result was an intellectual explosion that shook the foundation of the Establishment's power and propelled a new inquiry-driven trajectory.

By reducing the cost of reproducing and disseminating information Gutenberg moved ideas from the protective vault into a commercial environment that promoted its dissemination. Merchant printers created an information network by moving texts among themselves for reproduction, distribution, and profit.[4] That network, in turn, sparked the Reformation and spread the innovations of the Renaissance throughout Europe.

It would be 400 years before a new network technology disrupted the status quo once again. While Gutenberg's technology unlocked information and allowed it to travel, the journey remained physically constrained. From the beginning of time, distance had created walls between groups of human beings equal to or greater than the barriers created by the jealous guarding of scientific knowledge and intellectual expression. Early in the nineteenth century the steam locomotive powered through those walls to allow humanity to overcome geography's grip.

The iron horse dissolved the geographic isolation that had created independent, self-sufficient, local resource-based communities. By economically transporting high-bulk products, the railroad broke the connection between the location of the resource and the site of its consumption. In the process the railroad pulled both products and people off the land, feeding the pace of the Industrial Revolution. Towns that were once too far from rivers or the sea to engage in extensive commerce became hubs of activity tied together by ribbons of steel. The growth of the railroad transformed the landscape, remade cities, and disrupted the lives of millions.

As the railroad supplanted traditional pathways, the telegraph rode alongside. The two technologies experienced symbiotic growth as telegraph lines built along the railroad's rights-of-way carried messages that not only managed railroad activities but also introduced instantaneous communications into other aspects of life and business.

Whereas the railroad compressed distance, the telegraph condensed time. From the beginning of history, the fact that information moved physically meant that it moved slowly, limited to the same speed as human travel.[5] The telegraph separated the transfer of information from the transportation of hard copy. This virtualization of information further expanded the marketplace economy, brought forth unparalleled personal interaction, and laid the technological groundwork for the network that now defines the human experience.

As history's first electronic network, the telegraph was the internet of its time.[6] The only thing faster than a speeding locomotive, the telegraph controlled movement on the rails. But its impact was much more pervasive than managing train schedules. Information speeding faster than the wind made possible the creation of the Weather Bureau. News reports delivered from afar at lightning speed redefined both the nature of news and the news business. Electronic messages coordinated industrial production, created a new managerial class, and enabled the rise of powerful market-controlling corporations.[7]

The collective effect of these network revolutions was to gradually form an economy and a society of mass. The printing press created the first mass information economy. By one estimate, more books were printed in the first fifty years after Gutenberg's discovery than had been copied by all the scribes in Europe in the previous thousand years.[8]

The railroad then expedited the path to industrial mass production and a mass market. Before the railroad, production and processing activities were small operations distributed widely in locations adjacent to the raw materials. By making it possible to economically transport those raw resources to a central point for processing, the railroad fed ever-growing industrial complexes. Riding the rails in the opposite direction, the results of this mass production were delivered to a newly interconnected mass market.

The telegraph, and later its offspring, the telephone, tied together the new industrial activity. Mass production required coordination among the sources of raw materials, production facilities, and mass-market distributors. The telegraph was also the initiating force behind interconnected mass communications. When the newspaper publishers of New York banded together in 1846 to create the Associated Press, they were taking advantage of the telegraph's ability to collect information quickly from afar. In the process, they built the twentieth century's model of mass communications in which networks bring information to a central point for curation prior to its subsequent redistribution for mass consumption.

We may think we know the narrative of those earlier times, but our understanding is incomplete absent an appreciation of the linkage of that history with our lives today and, most important, our tomorrow. From a technological point of view, the earlier network breakthroughs are the roots from which grow today's "new" technologies. Sociologically, the changes driven by the earlier networks echo in the dislocations we experience today.

The evolution of network technology mimics the step-by-step natural evolution of living things. When Charles Darwin wrote that "it is the steady accumulation, through natural selection, of such differences . . . that gives rise to all the more important modifications of structure," he could have been describing technology as well as biology.[9]

New technology is an accretive process. While inventions are often described in terms of one person's inspiration, in reality they are typically a new assembly of accrued knowledge in a heretofore unrecognized manner for a previously unappreciated purpose. As we will see, Gutenberg's movable type was the coming together of a collection of known capabilities; the steam locomotive was a new way of dealing with a power understood since ancient times; and the concept of messaging through electromagnetic signals had been around for almost a century before Morse's "What hath God wrought."

"The process of technological development is like building a cathedral," Paul Baran observed. "Over the course of several hundred years: new people come along and each lays down a block on top of the old foundation, each saying, 'I built a cathedral.' Next month another block is placed atop the previous one. Then comes along a historian who asks, 'Who built this cathedral?' . . . But the reality is that each contribution has to follow onto previous work. Everything is tied to everything else."[10]

This is a book about our cathedrals, about the continuum of both additive and repetitive technological and sociological progress that lays the foundation for our future.

The first two network revolutions began with a centralizing force that expanded outward to create secondary and tertiary centralized hubs of network activity. Printing presses were originally centralized in academic and commercial centers. As printing expanded, it dispersed into multiple centers of activity. Similarly, the railroads brought people and products to a central point to be switched to the track leading to their destination. As the rails expanded geographically, switching also moved to satellite transfer points. The same topology held true for the railroad's partner, the telegraph, and then the telephone, with the switchboard performing the same function as the switching yard to route a call from one line to another.

It was the technological reiteration of the traditional pattern of networks. Historically, the initial impetus of a network was to create a central point from which its activity radiated. As it grew, that activity dissipated into decentralized hubs. The current network revolution is being driven by the ultimate expansion of network dispersal to further move activity away from central points to become fully distributed, ultimately right down to the individual.

While networks moved outward structurally, the economic activity they enabled moved in the opposite direction. Businesses seized on the network to build new centralized economic power. Rockefeller's Standard Oil, Carnegie Steel, Montgomery Ward and Sears, Roebuck mail order, Swift and Armour meatpacking, and others built centralized empires using the railroad and telegraph. Today we see the same pattern. As the distributed digital network pushes its functions outward, new businesses such as Google, Facebook, and Amazon ride that network to create new centralized powerhouses.

Whether historical or present day, the manner in which network-driven change develops is more redundant than revolutionary. Each of the network revolutions that on prior occasions redefined the nature of the human experience followed a similar pattern. First, a new technology breaks the ongoing incremental, linear evolution of the old technology by reformulating components that have been around for some time. Then the new nonlinear assembly is seized on by others to produce nonobvious results.

At the time of Gutenberg, for instance, the process of creating a book was going through such a linear evolution. The monks in the scriptoria were losing their monopoly on the reproduction of knowledge to the new business of commercial manuscript production. Yet, because it was merely the expansion of the existing high-cost, low-volume system, this logical, linear advancement had little transformational potential. It took Johannes Gutenberg's nonlinear thinking to create the nonobvious opportunities that accompanied an abundance of texts.

In the five and a half centuries following Gutenberg, his concept went through similar linear advancements—but it was still a process of putting stains on paper. It wasn't until Jeff Bezos took advantage of the arrival of a new network that the book was redefined to violate that incremental pattern. In the ultimate nonobvious innovation, Bezos's Amazon e-reader broke 550 years of precedent by separating the act of publishing from putting ink on paper.

The new network that allowed e-readers to upend half a millennium of ink on paper was itself the convergence of two previous examples of nonlinear thinking and nonobvious results. As we will see, the computing engine that powers an e-reader traces its lineage to the steam engine. And the network of binary electronic impulses that transports the contents to the device links to the telegraph.

Such nonobvious results create another socially relevant secondary effect. Because new networks dispatch old traditions, they trigger opposition from those who have grown comfortable with the old patterns. The stories we will visit demonstrate that a typical response to the effects of a new network consists of fear and resistance.

The Catholic Church, threatened by the unlocking of information it had always controlled to produce individual conclusions it could not control, attempted to suppress the output of printers.

The canal companies, stagecoach and haulage firms, tavern owners, and others who were bypassed by the speeding railroad used everything from political muscle to vigilantism to derail the iron horse.

Messages delivered by "lightning" became fodder for pastors to frighten the faithful about how it could only be black magic, while the U.S. Post Office resisted its opportunity to adopt a technology faster than the mail.

Ultimately, however, such responses proved to be the rearguard of a retreat in the face of those who saw opportunity in the new networks.

Martin Luther leapt from being an unknown monk to lead the Protestant Reformation by harnessing the power of the new printing shop network to deliver his message.

Chicago became the Second City of a growing nation, displacing St. Louis as the link to the West, because its city leaders aggressively embraced the railroad at the same time St. Louis was resisting it.

The newspaper business transformed itself from a collection of local political rags of limited scope into an electronically interconnected medium that made previously scarce information plentiful.

Though no small amount of social convulsion and dislocation accompanied the creation and promulgation of each of these network revolutions, their secondary effects nonetheless became comfortable and commonplace until each network was upset by the next new network. The network-driven upheaval of today is behaving similarly to upend the comfort that followed the standardization of what had been previous upheavals.

The centralized structures of the nineteenth and twentieth centuries in which networks assembled masses—of people, production, products, and information—is today yielding to a network through which forces move in the opposite direction to disaggregate and disperse activity.

The networks of history commanded the user to come to them: Come to the book. Come to the railhead. Come to the telegraph or telephone. The wirelessly distributed force of the new network does just the opposite. Now users command the network to come to them wherever they may be. It may seem a simple inversion, but its effect is this era's nonlinear, nonobvious result. It is nothing less than the transfer of the nexus of power from the network to the user.

The little-known and poorly understood secret of all the previous network revolutions was that the network was in control to create hierarchies and define activity. For the first time in history, the new network puts its user in control.

Behind this new network effect is the linkage between the technologies driving today's changes and the technologies of earlier networks. Our new network was spawned by the earlier technological breakthroughs.

Those breakthroughs began with Gutenberg—and Gutenberg's innovation was not simply a method for transferring ink to paper. His enduring insight was that for information to be automated and made "mass," it had to be disassembled into small units and then reassembled. It was a groundbreaking discovery about the interface between information and machines. The concept lives on at the heart of the internet's digital network, which breaks information into small packets for subsequent transmission and reassembly.

The success of the steam locomotive not only redefined production economies but also energized the belief that if engine power could replace muscle power, it could also be harnessed to replace gray matter. A spectator at the inaugural run of the first scheduled railroad, the British mathematician Charles Babbage, conceived of harnessing the same steam power to compute logarithmic tables. Thus began a developmental spiral that led from analog calculating engines to digital processors and ultimately to today's ubiquitous computing engine, the microprocessor.

By decoupling information from its physical form, the telegraph introduced the kind of virtual delivery that characterizes the internet. The dots and dashes of the telegraph's on-off signals are rearticulated by the binary signaling protocol of today's digital networks and processors.

Today's network echoes the networks of history in economic and behavioral aspects as well. The history of the networks that connect us is also the economic history of the world. When the economist Angus Maddison attempted in 2001 to estimate the growth of world gross domestic product (GDP) over time, the growth spurts he identified corresponded to the introduction of new network technologies.[11]

The first burst of sustained economic activity coincided with the era in which the printing press exponentially expanded the free flow of information. The book-fed Renaissance, Reformation, and the age of inquiry stimulated economic dynamism and recast the patterns of life.

The next spike in economic growth coincided with the railroad and the Industrial Revolution. As railroad track mileage and speed increased, so did economic development. The handmaiden to that expansion, the telegraph (and later the telephone), continued the network-driven economic spurt.

The relatively recent digital network era is consistent with the axiom that a new network stimulates new economic growth. The expansion of the digital economy also manifests another force that, while operational in the previous revolutions, has grown in significance: the increasing speeds of networks correlate with the acceleration of innovation and the pace of life.

From the dawn of the Christian calendar, it would take a millennium and a half for the printing press to emerge. From Gutenberg to the railroad and telegraph was only about 400 years. The period from the railroad and telegraph to the digital revolution was less than half that time.

We can see a direct relationship between the speed of a new network and the rate of innovative growth that speed stimulates. The acceleration of network speeds maps to the pace of technological change and the acceleration of economic and social change.

When information moved on horseback, it traveled at about four miles per hour.[12] Measured in terms of data throughput, this was about 0.03 bits per second.[13] The first network revolution, the movable-type printing press, increased the volume of data moved, but not its speed.

The railroad introduced speed into the equation. Moving five to ten times faster than animal power in its early iterations and later up to forty times faster, the iron horse accelerated the pace of life. The exponential growth of the railroad itself was a multiplier of the effect of its speed. In 1830 there were thirty miles of railroad track in the United States; by 1860 there were 30,000 miles of steel rails.[14] Before the end of that decade, 1869, the steam railroad had spanned the heretofore unbridgeable American continent, speeding change to the remotest regions.

The first electronic network, the telegraph, accelerated the flow of information yet again. At the time of the founding of the American republic it took twenty-two and a half days for information to move between New

York City and Charleston, South Carolina.[15] Four score years later, news of the dissolution of that republic in Charleston Harbor moved instantly to New York by telegraph. "The speed of the telegraph is about as wonderful a thing as the noble invention itself," observed *Scientific American* in 1852.[16]

A talented telegraph operator could transcribe at the rate of around three bits per second.[17] Making information instantaneously available everywhere at a speed 100 times faster than delivery by horse further hastened the pace of life and the rate of change.

Today's networks turn on the afterburners. Change is flying at us with gigabit connectivity (1 billion bits per second) and headed higher. That is more than 300 million times faster than the telegraph and 30 billion times faster than horseback.

As the velocity of the network increases, so does the speed with which it introduces change. In the process, twenty-first-century networks have destroyed the buffer that helped previous generations transition through change. Whereas previous new networks took years, if not generations, to mature, the current network revolution destroys old institutions and practices before it has ripened the capabilities necessary to replace them.

Data is speeding to us wherever we may be. On today's digital wireless networks not only does information fly fast, it flies to wherever the user wants it.[18] Giving the user, rather than the network, control to call forth the high-speed information he or she creates or consumes defines the era we are pioneering.

By definition, a network hub is a point where in-and-out activity occurs. Such hubs were formerly rail yards or switchboards, newsrooms or assembly plants. By emphasizing the virtual over the physical, our current network revolution not only makes information the preeminent product but also places in the hands of each person the ability to determine his or her own in-and-out patterns for that information. In these individual hubs users determine what they want to consume and with whom they want to connect. They create and distribute their own content as if they were the *New York Times* or NBC. And they perform their information-based jobs from locations of their choosing.[19]

In the pockets and purses of each individual hub is a powerful computer, colloquially called a "phone." The processing power of any one of these devices is greater than that once provided by huge machines locked in special rooms. These pocket computers speak a lingua franca that allows the information being delivered to be independent of both the network on which it travels and the device on which it is displayed.

The introduction of new technology that transforms the way in which we connect has thus come full circle. Gutenberg built the first information network by seeing information in terms of its smallest parts. Now the network has become the interconnected sum of its smallest parts—its users.

The act of publishing was previously centralized in the hands of those who owned the presses and the means of distributing their output. Today any individual can be an author and content creator with access to worldwide distribution.

Railroads pulled economic activity out of the hands of individual artisans and into massive industrial institutions. Now the skilled individual is returning to prominence, thanks to the ability to connect to a massive market without the need to be massive oneself.

The telegraph and telephone extended the user's reach but at the price of being tied to a wire coming through the wall. Now individuals can access wireless networks to deliver connectivity where the user is rather than where the wire is.

The mechanized productivity of information that began with Gutenberg, the power of engines that began with the harnessing of steam, and the binary transmission of information by electrons that began with the telegraph have all combined to create the third great network revolution. Accompanying this is the same kind of upheaval, opposition, opportunity, and stress that attended preceding network transformations.

History has been clear in the expectations it sets for our future. The innovations created by new networks topple old institutions and accelerate the pace of life. The demands of the new and the absence of traditional moorings generate frustration and bewilderment.

Like those who lived during earlier network changes, we are walking where there is no path. The stories of those who made the earlier paths are relevant not as "how-to" manuals but as landmarks. They are what navigators call a "back azimuth"—a known point to anchor and guide while we progress into the unknown.

The stories of previous network revolutions establish that "normal" is the institutionalization of what yesterday was inconceivable. They teach that in a time of technological turmoil, the greatest danger is not the turmoil itself but the attempt to cling to the comfortable ideas and institutions that remain from the last network revolution. Nonlinear thinking produced the technological change; its successful exploitation requires nonobvious innovation.

The challenge, of course, is the successful identification of the nonobvious. Later we will explore some of the issues created by our new network: the disappearance of privacy, the threat to jobs, the demands put on education, the rise of misinformation and its effects, both domestic and international.

But first we need the predicates of history.

How we connect defines who we are. The story of the human experience is how new means of communicating created new economic and social institutions. What follows are the stories of those connections. They are the history of our future.

Part II

Predicates

Two

The Original Information Revolution

The crisp fall wind rustled the pages nailed to the door of Wittenberg Castle Church. It was All Hallows' Eve, October 31, 1517. The following day the town that took its name from the castle would be jammed for All Saints' Day.

The pages that had just been posted on the church door were the work of a monk from the local Augustinian monastery. The church door was the town bulletin board; such postings were a way of triggering a theological discussion at the town's new university and its monastery. This time, however, something different happened. Rather than simply receiving constrained consideration among the town's intelligentsia, the monk's posting leapt from the church door and raced across Europe's landscape.

The monk responsible for the posting was Martin Luther, and his thoughts spread owing to a new technology that created a new information network.

The pages on the church door contained a set of ninety-five theses in which Luther questioned the role of the Catholic Church as the only intermediary with the Almighty. It was heresy.

Such heresy was not unheard of; it had just gone unnoticed. For centuries, men of the cloth had proposed new theological constructions. These ideas, however, remained limited to the locale of their proponent's preaching. The Catholic Church, after all, controlled the production of books by which the concepts could be propagated.

By the time Luther put forth his ideas, the Church's information monopoly had been broken. Luther's good fortune was that his thoughts coincided with the spread of a new commercial technology for the reproduction and distribution of ideas. Printing with movable type had eliminated the choke hold on the flow of ideas that manuscript scribes and their clerical superiors had previously exercised.

The document posted on the church door leapt to the local merchant printers and then to the presses of Europe. Tacked to the church door, Luther's ideas had merely rustled in the wind. Set in type, they blew across Europe with a gale force.

Printing added unprecedented velocity to the spread of information. In the process, the network of printers made the previously unknown monk from Wittenberg the world's first mass-media evangelist.

When Ideas Travel

Situated in the northeastern quadrant of Germany, sixteenth-century Wittenberg was a center of political and commercial activity. The new university also made the town an intellectual center. Thanks to the university's demand for texts, the relatively new technology of printing with movable type had gained a foothold in the town. In late 1517 and early 1518, Martin Luther's thoughts combined with the new technology, launching what today we refer to as the Reformation.

In his Ninety-Five Theses, Luther challenged the prevailing religious orthodoxy. Why, he argued, should an individual be able to reach God only through the Church? If we are all children of God, could not an offspring reach the Father without the intervention of clerics?

When word of the monk's heresy reached the Holy See in Rome, the corpulent and conniving Pope Leo X was unperturbed. Historical precedent abounded that such challenges remained isolated events and ultimately died out.[1] The pontiff dismissed the Wittenberg posting as "the ramblings of a drunken German [who will] think differently when he sobers up."[2] Six months later, the stone-cold-sober monk had not changed his mind, and his ideas had taken flight to permeate the pope's entire domain.

Called to account for his actions, the monk dissembled. "It is a mystery to me how my theses . . . were spread to so many places," he explained in a letter to the pope.[3] It was a disingenuous denial. The monk may have been surprised at the speed with which his ideas spread, but there was little mystery as to *how* they spread.[4] Soon much of Europe was drunk on Luther's idea that individuals and not the Church controlled their relationship with God.

Rather than withering as the wind and weather attacked the pages hanging on the church door—yet another idea constrained by local factors—the economic opportunism of the network of new printing establishments accelerated Luther's ideas to escape velocity. Soon additional commentary from his pen poured out of printing establishments at such a rate that Luther's screeds were nicknamed *Flugschriften* or "flying writings."[5]

Luther was no stranger to the network potential of the printing press. Only a few years earlier, in 1508, a commercial printer had set up shop within the walls of the monastery where he lived. Such proximity no doubt provided familiarity with the technology and insight into its potential.[6]

Luther had taken advantage of that familiarity even before posting the Ninety-Five Theses on the church door. A year earlier the monk had convinced the printer to reproduce the sermon of another German preacher, to which Luther had added his own introductory comments.[7] It was a trial run for how the printing press could propagate his thoughts.

One of the most interesting twists of history is how Martin Luther's theological revolution was triggered by his revulsion to an economic mainstay of the printing business. The technology that turned Luther's ideas into "flying writings" was also the technology that very profitably produced letters of indulgence for the Church. It was the practice of selling such indulgences that drove the monk to challenge the institution of his faith.

An indulgence was a medieval "get-out-of-jail free" card. Since the Church controlled each person's relationship with God, getting right with the Church would get you right with God. One way to get right with the Church was with cash.

The Catholic Church badly needed cash in the early sixteenth century. Pope Leo X—the first of the Medici popes and the last pontiff who was not a priest—had reportedly remarked, "Since God has given us the Papacy, let us enjoy it."[8] Two years of such enjoyment was all it took for the new pope to deplete the Vatican's treasury. Yet demands on that treasury continued, not the least of which was the need to finish building St. Peter's Basilica.

The pope's liquidity crisis happened to coincide with the archbishop of Mainz's desire to expand his power by adding a third bishopric to his realm. While it was technically against church law to have a third bishopric, the pope could, of course, provide dispensation from the rules. A deal was consummated. The archbishop enriched the Vatican's coffers by 10,000 ducats (an amount allegedly made holy by using the same base number as the Ten Commandments), and the pope granted dispensation.[9]

To raise the necessary funds, the archbishop sold letters of indulgence to his flock. The indulgences were a well-established money-raising scheme that dated to the Crusades whereby the Church could fill its coffers by allowing individuals to purchase their way out of Purgatory. This time, however, the archbishop improved on that formula. To ensure his fundraising success, the archbishop expanded the power of the certificate retroactively. It became possible not only to purchase protection for yourself but also to purchase a release from Purgatory for the deceased.

"Can you hear your dead relatives screaming out in Purgatory while you fiddle away your money?" went the sales pitch. One particularly theatrical monk even devised a little ditty to promote his sales activities:

When a coin in the coffer rings,
a soul from Purgatory springs.[10]

That kind of powerful sales pitch meant that a large supply of indulgences would be needed. Such orders for indulgences were a godsend to commercial printers. Long runs of a simple document were a printer's most profitable work. Best of all, the entire output sold at once to a single buyer. These indulgences were produced in massive amounts; by one estimate, the number of indulgences printed throughout Europe for this and other money-raising campaigns climbed into the hundreds of thousands.[11]

But what was manna from heaven for the printers was an anathema to Martin Luther. In the quiet of the monastery and the intellectual stimulation of the classroom, Luther had been privately exploring the concept of reaching God without the intervention of the Church.

The sale of indulgences caused Luther to go public with his thoughts. Appalled by the profiteering of a church that sold God's grace for cash, Luther first went through appropriate channels and wrote to the archbishop. Of course, it was the archbishop who was using indulgence revenue to expand his realm. Shut out, the monk nailed his theses to the church door.

Attaching a challenge to the Church on the door of its house of worship was less a symbolic confrontation than a simple posting on a community bulletin board. The church door was pockmarked with holes from other such postings placed there as a means of beginning a discussion among the local academic community.[12]

It was no wonder that the pope was unconcerned. Since the theses were written in Latin, only the well educated would understand them. Furthermore, exposure to the heresy would be physically limited to those in Wittenberg.

This time, however, the ink hit the press. Within fifteen days Luther's theses, translated from Latin into the German language of the masses, were available in every part of Germany.[13]

The monk had met his medium.

By the beginning of the sixteenth century there were commercial printers in sixty German towns. Elsewhere throughout Europe, similar infrastructure was being built.[14] Like any other commercial entity, these printers were constantly searching for new products to produce and sell. Modern history is replete with examples of newspapers churning up controversy to sell their product. The incentive was no different in the sixteenth century. Luther's ideas fit the printers' demand equation perfectly.[15]

The no-longer-obscure monk and his work became a profit machine for commercial printers. Following up on the best-selling flyers describing the Ninety-Five Theses, Luther sent to the presses the "Sermon on Grace and Indulgences," further explaining his ideas. This direct-to-the-presses attack on indulgences became Luther's biggest hit, being reprinted in fourteen editions in 1518 and another eight in 1519–20.[16]

Having found both his voice and his vehicle, Luther became prolific. The year after his theses, he published eighteen new works, most in the language of the people. By writing in German rather than in clerical Latin, Luther greatly expanded the audience for his message and thus for the printers' output. Soon he was even releasing a Bible written entirely in the native tongue, so the people could interpret God's word for themselves. The first edition of the New Testament in German (1522) sold out in a matter of weeks. In the following two years eighty editions were published throughout Germany.[17]

"The indulgences market collapsed like a popped dot-com," one modern observer commented.[18] The printers for whom the Church had been a mainstay of business could not have cared less, however. By one account, one-third of all the books printed in Germany from 1518 to 1525 were the product of Martin Luther's pen.[19]

Reassembling the Old to Create the New

It is a wonder that the printing press that gave Luther his medium ever came to be.

For centuries, mankind had been on the precipice of printing. The components necessary for the movable-type printing press were all present in one form or another in the fifteenth century. Their assembly, however, was a medieval Rubik's Cube of permutations and combinations.

The screw press was not new. As early as the Roman era, presses had been used to squeeze grapes and olives for their juice.

The assembling of individual characters, cut in reverse and inked to create an impression, had been used by the Chinese since the eleventh century. While the idea had not reached Europe, it was nonetheless clearly within human comprehension.

Inexpensive paper had come to Europe in the twelfth century from Arab countries, where the techniques for its production had been learned from Chinese prisoners four centuries earlier.

Ink had been around since the development of writing. Mixing carbon soot with various liquids to produce a mark was far from revolutionary.

While all of these preexisting capabilities had never been brought together, the printing press was more than their simple amalgam. Solving one part of the puzzle had the effect of opening the door to another problem. Conceptualizing how these basic techniques could work in harmony to enable mechanized printing was a sizable intellectual hurdle. Actually accomplishing that synergy was made infinitely more difficult by the need to adjust each technique so it would work with the others.

The press had to do more than apply a squeezing pressure. The pressure had to be delivered uniformly throughout; otherwise some text would be lighter or less sharp than others.

The individual characters had to be consistent in size and shape. Yes, the Chinese were the first to cut their characters into molds (first for clay characters and eventually for metal type), but the results were not uniform, and each character emerged from the mold with its own idiosyncrasies.

Most paper was too thin and too absorbent for printing on both sides. Imagine trying to print consistently on something similar to bathroom tissue!

The water-based ink ran off the metal type to which it needed to adhere.

The compounded challenge of solving the base problem of the press, paper, ink, and type and then making them all work in concert was the decade-plus quest of a German goldsmith named Johannes Gensfleisch Gutenberg.[20] The magnitude of his challenge and the revolutionary concepts he was developing are revealed by the records of a 1439 lawsuit.

Gutenberg's concept was so bold, and the process to accomplish it so revolutionary, that in 1438 he and three others entered into a contract that swore each of them to secrecy. What they called the "secret art" was so special that the contract further provided that should one of the partners die, his heirs would have no access to the secrets.

Within a year, the plague had claimed one of the partners.

As the partners had feared, the deceased's siblings tried to force their way into knowledge about and an ownership share of the "secret art." As part of this effort they filed suit. A trial was conducted to determine whether the brothers of the dead partner could claim his share. The records of the trial forge the link between Gutenberg and the discovery of the movable-type printing press.[21] The records also display the extent to which Gutenberg went to protect the results of his work.

While the witnesses tried to obfuscate exactly what was going on, it appears Gutenberg's secret art was his early effort to resolve the interlinked issues necessary for a movable-type printing press. It was a secret so profound that when the partnership was threatened by the legal action, Gutenberg ordered the destruction of his hard work lest the prototypes fall into the wrong hands.

Johannes Gutenberg had been born into the emerging medieval middle class of craftsmen. His father was a goldsmith, a "Companion of the Mint" in Mainz, Germany, where he produced coinage. Johannes followed in his father's footsteps to learn the art of working with metals.

The son was a skilled craftsman. He was also an entrepreneur in an age when entrepreneurs were building a new middle class. One of his many business activities—the buying and selling of wine—put him in proximity to the screw press used for squeezing wine grapes. A failed business undertaking, which probably also used a screw press, gave him further exposure to working with molten metal and produced the partnership that ended up in court.

The original business idea of the partners had nothing to do with printing. It was a simple plan to exploit the traditions of the Catholic Church. Because the holy relics of the Church cemented the clergy's links to the saints and even to Christ, the Church would unveil the relics on special occasions. The faithful were told that gazing upon those artifacts connected the viewer to the powers they represented. As a result, the unveiling of such relics became grand occasions that stimulated great pilgrimages.

One of the greatest pilgrimages of medieval times was to Aachen (about 160 miles northwest of present-day Strasbourg, France) to see the remains of the sainted Charlemagne, the swaddling clothes of the baby Jesus, the loincloth of the crucified Christ, and other purported memorabilia.[22] The Aachen relics were exhibited once every seven years. For the two weeks of their display, upward of 10,000 pilgrims per day would gather to gaze upon them from afar. During the 1432 pilgrimage the belief developed that if a convex mirror was held so as to capture the relics' reflected image, it would absorb and store their radiant powers. To Gutenberg, that belief was an economic opportunity.

Living at the time in St. Arbogast, outside Strasbourg, Gutenberg developed a plan to manufacture and sell such mirrors for the 1439 pilgrimage. Making a mirror was difficult, making a convex mirror more so. In these challenges, however, Gutenberg was served well by his experience in the family trade of working with metals. To raise the capital necessary to purchase and fabricate the raw materials, he took in three partners.

Unfortunately, as happens to so many business plans, the fates conspired against its success. The 1439 pilgrimage was canceled because of the spread of the plague. Gutenberg and his partners were left with an inventory of raw material and partially completed trinkets.

The partnership, however, held together. Whether the work on the trinket mirrors opened some insight into what became movable-type printing can only be speculated (it did, after all, involve bending the shape with a press and creating a mirror using molten metal, both skills important to making movable type practicable). Nonetheless, something happened to encourage the partners to enter into a new five-year contract and invest further funds. It was that contract that was being adjudicated in order to protect the secret art.

When the court found in favor of Gutenberg, it was a two-pronged victory. First, the interlopers were kept out and Gutenberg's group was able to buy back the deceased partner's share at cost. Second, any of the witnesses who may have known what Gutenberg was up to held their tongues and spoke only in the most circumspect manner. Testimony documented, nonetheless, that Gutenberg was doing something that involved a press, smelting of metal, "formes," and "four pieces" held together by "two screws."[23]

Gutenberg fulfilled his partnership obligation and stayed in Strasbourg for its five-year term. During that period the fits and starts of his exploration continued, but the great breakthrough did not occur. At the end of the agreement he left town.

By 1448, Gutenberg was back in Mainz, the city of his birth, about two days north on the Rhine River from Strasbourg. For over a decade he had been patiently playing with the Rubik's Cube of diverse pieces necessary for the mastery of movable type. In Mainz it all came together.

What passed for printing in fifteenth-century Europe was the use of blocks of wood into which the letters and pictures of an entire page had been laboriously carved. Blotting this woodcut against a piece of paper produced a complete page.[24] The problem with the process, however, was the labor-intensive carving of the woodblock and the imperfect reproduction that resulted from blotting.

The idea of seeing the page not in its entirety but as a collection of smaller pieces of information was an intellectual breakthrough in Western thought.[25] Beyond that breakthrough, however, the problem remained how to make a collection of identical type and lash the pieces together to produce the same result as a woodcut.

Carving a relief letter on the head of a punch was a well-known method used for embossing the leather covers of manuscripts and making impressions in the dies that struck coins. Because the letters were each carved by hand, however, they were not uniform. The labor-intensive handcrafting of each letter also limited the scale at which the letters could be produced.

Taking such a punch and driving it into a softer metal would produce a shape, called a matrix, which could then be filled with molten metal to produce a copy of the shape. This could have solved the scaling problem but for the fact that metal contracts as it cools, and thus each letter from the matrix came out slightly different. Further compounding the complexities of using a matrix was that small variations in its creation were reproduced in its product. The angle at which the punch hit the matrix and the force of the blow that set the matrix's impression affected the shape of the type it produced, differentiating it from the desired consistency.[26]

Even if these problems were solved, Gutenberg still had to come up with solutions to such thorny issues as the varying widths of letters. Clearly, the letter *m* is wider than the letter *i*, and a capital *M* is wider than a lowercase *m*. Yet when the letters were all bound together, the spaces between the letters had to appear uniform.

No wonder Gutenberg had been laboring against these challenges for a decade or more of trial and error. And creating the type itself was just part of the problem. Gutenberg struggled to achieve other breakthroughs necessary for his idea to work.

The evolution of medieval underwear helped solve one of Gutenberg's related problems.[27] Wool undergarments had clothed the populace since the beginning of haberdashery. Around the twelfth century, however, linen was substituted for wool. To its wearers, that in and of itself must have seemed a great leap forward. The shift had the unintended effect, however, of initiating a rag trade of discarded linen, which became a plentiful and low-cost source of raw material for paper.[28]

A decline in the price and increase in the quality of paper was helpful, yet the physical qualities of the paper presented their own problems. Because of its fragile consistency, the surface of early paper had to be treated to make it harder and more impervious to ink leaking through to the other side. That hardening, while fine for the scribes' quills and ink, inhibited the paper from accepting the ink from the press. More trial and error revealed that moist paper would evenly absorb the ink from the type. The problem then became determining the correct amount of moisture and developing a process for both moistening the pages before printing and drying each page and its inked impression afterward.[29]

The solution: the dampening of every other sheet in a stack of paper before putting the whole lot under a press would allow the moisture to leach from wet to dry pages. Like so many of Gutenberg's "solutions," this advance led to a new set of challenges. Paper could stay moist for only a finite period before it began disintegrating and losing its strength. The length of time it could stay moist without damage changed constantly as a result of changes in the barometric pressure outside the printing shop.

Then there was the problem of the ink. The ink used by scribes was plentiful, but the viscosity that enabled it to flow smoothly out of a quill meant it similarly ran off the metal type. Once again, painstaking trial and error afforded a solution. What Gutenberg needed was an oily ink that would stick. He found his solution in the varnish-like paint used by Flemish artists. The mixture of lampblack soot, boiled linseed oil, heated lead, and copper oxide stuck to the type.

Getting that ink onto the type was yet another challenge requiring a new solution. How could the new ink be applied so that it didn't fill the holes in letters such as *b* or *e*? For this Gutenberg developed ink balls. Looking like half a grapefruit stuffed with wool or hair on a stick, the ink balls were doused in a tray of ink and then wiped over the type to deliver an even coating that adhered only to the top of the type pieces and didn't run down into the holes.[30]

What had previously passed for "printing" from woodcuts was actually more like blotting. Sharp images, however, required pressure. The technology of the screw press may not have been new, but its application to printing created new problems. A grape on the outside of a wine press doesn't need to be squeezed at exactly the same pressure as a grape closer to the center of the press—but the last letter on the last line of a printed text must receive the same amount of pressure as all other letters or else its imprinted image will be lighter or darker, crisper or fuzzier. Similarly, while it makes no difference if the plate pressing down on a grape slides a bit laterally, even the slightest such movement on a printing press will smear the type.

Gutenberg's tinkering with the pressing mechanism resulted in an upper plate that descended evenly with equal pressure throughout. It was yet another trial-and-error improvement on an existing technology.

None of the aforementioned challenges were trivial. The technique for reproducing identical pieces of type and holding them together to make a page, however, was Gutenberg's greatest challenge.

The contraction of metal as it cooled had prevented the production of identical type and was solved by adding the element antimony to the molten mixture of tin and lead.[31] Antimony expands as it cools; pressing against the walls of the matrix rather than contracting produces a uniform result from each pouring. But antimony is highly toxic. Not only was Gutenberg working with molten metals heated to over 600 degrees Fahrenheit, but the vapors coming from the liquid were also poisonous.

While each letter was now identical to the others of its kind, there remained the challenge of the differing widths of letters, as well as producing a consistent-length base on which the letters stood. As anyone who has tried to shorten the leg of a table can attest, trying to file the stem on which each letter stood to a common length was an impossible undertaking. Infinitesimal variation in the height of a letter would produce an impression different from that of the letter next to it.

Gutenberg found the solution to both these problems in a four-piece handheld mold. Two L-shaped molded pieces fit together and the matrix with the letter's impression would slide into a slot in the bottom. The width of the stalk of the letter could be adjusted to the characteristics of the letter based on the width of the matrix. The pieces were held together by a metal loop, the molten metal was poured, the loop was released, the two L-shaped sides fell apart, and there stood a uniform piece of type. Recalling the lawsuit testimony about "four pieces" held together by "two screws," Gutenberg's effort appears to have been an early attempt at a typeface mold.

With expanding metal, a constant impression of the letter being cast, and a four-part mold to ensure standard size, Gutenberg had solved one of his most vexing problems. The problems that remained, however, were equally aggravating. Having successfully created the smallest unit of information on the page, he now had to reassemble the pieces and hold them together as a single unit.

The 1439 court testimony also discussed something called "formes." The type had to be fitted into a frame that tightly bound the letters together. This wooden frame was solid on three sides while the fourth side opened to allow the type to slide in. Blanks of differing widths were used to align the ends of the lines. When the form was filled, the fourth side was attached, and small thin pieces of wood were inserted to act like shims and hold the pieces tight. The result was a solid mass—the equivalent of a woodblock—from which to print. Unlike the woodblock, however, the plate was a solid mass only in its effect; in reality the plate was the temporary assembly of the smallest usable components of information.

Around 1450, Gutenberg's presses began producing finished products.[32] The first commercial product was probably a schoolbook, the twenty-eight-page Latin text *Ars grammatica* by Aelius Donatus.

Success!

It is worthwhile pausing at this point to savor Gutenberg's success.

Imagine the exultation and celebration that must have gripped Johannes Gutenberg as his first printed book was bound!

More than a decade in development, Gutenberg's understanding that a page of information was the sum of its parts had required a "secret art" to both discover a revolutionary new process and find the means of adjusting a seemingly endless number of variables into harmonious production.

Now it was done. Success had been achieved in twenty-eight pages of Latin grammar instruction.

The Western world had never before seen the rapid production of hundreds of perfect-quality pages, each one identical to the others. It was a moment to be savored, a decade-long quest with a transformative result.

Unfortunately, the exultation would be short-lived.

Other mass-market documents flowed from Gutenberg's printing shop. The earliest dated work was a papal indulgence of 1454.[33] Having spent more than a decade perfecting his technique, however, Gutenberg, it would appear, was not satisfied with such run-of-the-mill products. He wanted a monument. Today we call that monument the Gutenberg Bible.

It would be his downfall.

Economically, printing was a capital-intensive undertaking. Large amounts of raw materials had to be purchased and held in inventory awaiting processing into a finished product, which then waited to be sold. This was especially true of the proposed Bible, a voluminous 1,275 pages. For a production run of around 175 copies (135 on paper, 40 on calfskin vellum), the project would require 5,000 calfskins for the vellum copies and more than 250,000 sheets of paper for the 135 less expensive versions.[34] These investments were on top of the type that had to be produced (for which a new style was especially developed), ink that had to be bought, equipment that had to be built (a second printing shop was set up), and employees who had to be hired and trained.

To fund this undertaking Gutenberg turned to the Mainz businessman Johannes Fust. Fust had already lent him 800 gulden for the original printing shop. For a second 800-gulden loan to finance the Bible, Fust wanted increased security: Gutenberg's Bible workshop and its equipment. If Gutenberg did not repay the loan, the printing shop would be Fust's collateral.

In 1455, just as the Bibles approached completion—and the generation of revenue—Fust called his notes. Together with interest they totaled 2,026 gulden. Of course, Gutenberg could not pay; his assets were tied up in the about-to-be-sold books and the innovative technology to which he had devoted more than a decade of his life.

Johannes Fust became the new owner of the world's largest inventory of Bibles and the revolutionary technology by which they were produced. He and his son-in-law (Gutenberg's assistant, Peter Schöffer) took over the business.

After such shrewd—some might say dastardly—behavior, Fust perhaps got his due. There is an often repeated (perhaps apocryphal) story about Fust's efforts to sell his Bibles. The best market for his product was the city with more universities and more students than any other in Europe. Off to Paris went Fust and his ill-won gains. He reportedly sold one vellum copy to the king and another to the archbishop of Paris. Paper editions were sold to the lesser clergy and common folk.

Apparently, however, Fust did not disclose how his Bibles had been created. When the archbishop, proud of his purchase, showed it to the king, they discovered to their amazement that except for the hand-painted illuminations, every page of their two books was identical. The Confrérie des Libraries, Relieurs, Elumineurs, Escrivans et Parcheminiers—the book producers' guild—was called to give its opinion.

Professional inspection confirmed the books were not the work of scribes. The craftsmen whose skills were threatened by the new technology opined that such perfection could not have been achieved other than by dubious means. The Church declared such perfect copies could only be the work of the Devil.

Johannes Fust was accused of being a heretic. Threatened with being burned at the stake, he fled.[35]

The Original Information Revolution

The marvel of Johannes Gutenberg's persistent pursuit of his vision is exceeded only by its impact. By unlocking the free flow of information, Gutenberg's breakthrough was the open sesame to discovery, innovation, and the expansion of knowledge that enabled every scientific and technological advance that followed. As ideas flourished, they began to procreate and produce even newer ideas and innovations.

For centuries, the priestly and the powerful had had a Janus-like impact on the flow of information. While the monks and friars in scriptoria reproduced and preserved knowledge, access to such information was largely confined to the libraries of abbeys and castles, there to serve the purposes of its owners. Such a monopoly on information helped the nobility and the priesthood protect their positions, a reality that was challenged by the network of printing press operators that sprang up across Europe.

Because the cost of reproducing information was dramatically lowered, information became more plentiful. By one estimate, a printed book was 300 times less expensive to produce than a scribe's manuscript.[36] As access to ideas became more open, competition among ideas flourished and more minds were stimulated. The newly engaged then challenged one another with debate and dissent, which produced additional volumes. It was a self-perpetuating cycle that continually expanded the breadth of knowledge.

As a result of printing, Western civilization emerged into a golden era of discovery, innovation, and expansion.

Information traveled. Books, pamphlets, and flyers in backpacks and saddlebags spread ideas wherever people wandered. As the information moved about, the books of one printer provisioned the presses of others. When a printer in one town reprinted the books of a printer from afar to supply his unquenchable thirst for new items to sell, the cycle of innovation and new ideas expanded even further.

Knowledge gained permanence. When a single book is hand-copied, its information is fragile and potentially fleeting. The frequency of disasters such as war, disease, and famine both destroyed the hard copy of the knowledge and decimated those generations that might pass on the knowledge. Hundreds or thousands of copies of the same product, dispersed across the landscape, increased the probability the ideas would endure—they became resilient.

Gutenberg's gift to history is exemplified in the Renaissance.[37]

When Rome fell in 476, the trade-based economies within its dominion collapsed. Even the great city of Rome itself atrophied. As the commercial activity of cities shrank, so did the literacy and inquiry such activity supported.

It was the monks, friars, and their noble benefactors who preserved knowledge through their scriptoria and libraries. Little wonder, then, that the ideas that were hand-copied into vellum-paged books were ideas that seldom rocked the boat for those who supported the monks' activities. Printing upset that stability by creating a private economic incentive to seek out new ideas as the basis for the production and sale of new printed products. The pursuit of profit transformed those who reproduced information from gatekeepers that controlled access to it to widespread disseminators of it.

The result was the original information revolution. As a printer's run of hundreds of copies of the same book increased the odds of survival for that information, it also scattered the thoughts as seeds.

The printing press created a world that encouraged dissemination, discourse, and debate, and became an idea-generating process in and of itself. What we know as the scientific method—the formulation of a hypothesis and its empirical testing—spread to a wider constituency, as printing emerged as an argumentative and discursive platform.

Within half a century of Gutenberg's breakthrough, printing establishments had been founded in every major European city. Printing spread like a virus, slowly at first and then at an accelerating pace. It was an information explosion. For more than a thousand years, scribes had labored to produce manuscripts; in the first fifty years after the discovery in Mainz, more books were printed than had been produced in a millennium.[38] There was an insurgency of ideas as information and the knowledge it created spread across all aspects of life.

The Renaissance, which had its first stirrings in Northern Italy in the mid-fourteenth century, was spread by Gutenberg's mid-fifteenth-century invention. Lost and decayed Latin manuscripts were rescued and reintroduced. The humanistic teachings of Greece and Rome spurred further intellectual exchange.

Had it not been for the distributive powers of the printing press, however, this intellectual flowering might have remained isolated in Northern

Italy, or at least slowed in its progression. As one observer has noted, the Renaissance was "a period of decompartmentalization: a period which broke down the barriers that kept things in order—but also apart—during the Middle Ages."[39] The inexpensive reproduction of information was the departitioning vehicle.

Far from today's image of the Renaissance as an almost magical time, the period must have been anything but magical for those living through its changes. The discovery and rebirth that gave the period its name were, by their very nature, destabilizing to the patterns of everyday life. Gutenberg may have helped expand inquiry, but that inquiry undermined the stability and security of the status quo for everyone, regardless of their station.

For the first time, mass media appeared. Change—creating new ideas—became the bread and butter of printed media. In their search for new material, printers made the Greek and Roman classics more accessible. When Renaissance scholars explored the wisdom in the classics and propounded their own ideas, they stimulated a noisy, media-driven, humanistic debate about philosophy, science, and art.

As ideas built on ideas, they created further commercial opportunities, which in turn stimulated more debate. Printers in search of new content encouraged new texts containing new ideas that prompted not only a change in thinking about the humanities and science but also a reshaping of commerce.

Double-entry bookkeeping, a technique Venetian merchants and bankers had exploited to great advantage, became known to the rest of the world in 1494 when the Franciscan monk Luca Pacioli included it in his book, *Summa de arithmetica, geometria, proportioni, et proportionalita* (Everything about Arithmetic, Geometry, and Proportions). By explaining money as a mathematical proposition (assets = liabilities + equity), Pacioli's book explained to the world how money could be "created" on paper. The results financed exploration, innovation, and expansion.[40]

The printed word's stabilizing and distributing effect—improvement by revision—advanced the tools of trade beyond bookkeeping. As ideas and experiences regarding ship construction were exchanged and revised in print, shipbuilders produced larger vessels capable of traveling greater distances. The same process captured and continually updated cartographic knowledge to expand the routes those ships could travel.

The masters of those great ships devoured the new information contained in printed texts. Shortly after he arrived in Spain, Christopher Columbus bought several new volumes on geography. His copy of the first printed edition of *The Image of the World* included theories about the extent of the oceans. When pressed by the royal court to justify his concepts, Columbus produced this and other printed volumes as corroborating evidence.[41]

Many of the items we take for granted today originated as products of the printed word. As shipbuilding and navigation improved, for instance, English woolens began to arrive in Italy to undercut the domestic market. Italian businesses, in response, looked to their specialized strengths to develop an alternative market of high-quality goods that continues today.

Consumers accessing the newly available books often discovered they were farsighted and needed glasses. This new demand for optical lenses in turn expanded experimentation in optical physics, which led to the development of the microscope and the discovery of cellular biology.[42]

Similarly, the kind of statistical analysis that permeates our lives, from television ratings to political polling and insurance premiums, began with the sixteenth-century publication of census information.[43] Even Latin, formerly the domain of contracts and of the Church, began its trek toward the mortuary. As printers published in the vernacular to attract the widest audience, understanding Latin was no longer the prerequisite to accessing information in texts.[44]

The printed word also changed the relationship between the governing and the governed. The reality that any person could potentially access scientific fact inexorably led to the idea that people other than the king and his nobles could determine truth.[45] And just as scientific debates in print spurred further investigation, publishing also spurred political discourse. But printing added a further component to political life: it made decisions precise, permanent, and pervasive. Laws printed in permanent volumes were distributed among and accessible by the governed, enabling a codified, precedent-based legal system. Even more important, such hard-copy records gave the people the ability to police the laws' implementation, making the granting of rights virtually irrevocable.[46]

Down with the New!

Precisely because of the absence of an information tool such as the printing press, there was no universally accessible historical record to offer perspective on the experience of change. A natural result for its opponents, therefore, was to go after the force driving the change. The new technology and the flood of books it produced created for some a problem in need of a solution.

Early in the life of the printing press, the Establishment lauded its benefits. The archbishop of Gutenberg's hometown proclaimed the printed Bible a "divine art."[47] Not only was the Good Book itself propagated by printing, but the presses also inexpensively churned out an endless supply of money-raising indulgences. The benefits of low-cost common texts also promised a new consistency in Church documents, replacing scribal errors in hand-copied manuscripts and local variations on doctrine with a uniform text.[48] Poor and remote parishes would now have the same tools and rules as the great cathedrals.

The destabilizing impact of uncontrolled access to information soon began to challenge both Church and state. Troubled leaders in both institutions sought to protect their prerogatives against the insurgency of information. In 1475, just twenty-five years after Gutenberg's first commercial product, the pope authorized the University of Cologne, located less than ninety miles from Mainz, to censor both books and their readers. Eleven years later the archbishop of Mainz, who had earlier praised printing, ordered the inspection of all books. Ultimately, he decreed that anyone who published a book without prior approval would be excommunicated. Similar strictures were imposed by other authorities throughout Europe. Finally, in 1515, Pope Leo X forbade the printing of any book without the prepublication consent of Church authorities.[49]

It was the vicar of Croydon who most truthfully encapsulated the Establishment's concern with his warning: "We must root out printing or printing will root out us."[50]

But the rooting had already begun. The raucous reproduction of ideas continued unabated. The rules about prepublication clearance did nothing to stop the dissemination of Luther's writings. While the number of banned books continued to grow, and the secular state soon joined the Church in policing publications, the bureaucracies of the enforcers were no match for the printers' economic imperative. The cavalcade of printed ideas continued.

Concern about the impact of printing did not recede with time. Almost a hundred years after Gutenberg's breakthrough, the Swiss scholar Conrad Gesner set out to catalog all the books that had been published. Amazingly, in his preface to *Bibliotheca universalis* (1545), Gesner warned that something had to be done about the "confusing and harmful abundance of books."[51]

After another hundred years, concern about the effect of unfettered information still had not abated. As the French scholar Adrien Baillet warned in 1685, "We have reason to fear that the magnitude of books which grows every day in a prodigious fashion will make the following centuries fall into a state as barbarous as that of the centuries that followed the fall of the Roman Empire."[52]

The printing revolution demonstrated a reality that persists today: a new network technology produces upheaval long before it produces stability.

The effect of the new technology on jobs also set a pattern that would echo in subsequent network revolutions.

The employment in question this time was that of the monastic scribes. The abbot of Sponheim, Johannes Trithemius, came to the tradition's defense in *De laude scriptorum* ("In Praise of Scribes") in 1492. Trithemius contrasted the virtuous culture of scribes with the soulless printing technology. "In no other business of the active life does the monk come closer to perfection than when *caritas* drives him to keep watch in the night copying the divine scriptures."[53] Brother Trithemius also warned about the practical aspects of printing. Text printed on paper could survive, he calculated, for 200 years, while a monk's script on vellum would last at least a thousand years. He also argued that the scribes were more conscientious about spelling and textual correctness, thus making their product more reliable than the work of printers.

Trithemius's "In Praise of Scribes" was a tour-de-force defense of his monks and the practices displaced by the printing press. He was apparently so proud of his work that to ensure its widest circulation, Brother Trithemius had his paean to scribes set in type and printed!

Creating Today

If, observing the entire course of human history, one were to draw a line between "then" and "now," the line would pass through the printing shop of Johannes Gutenberg.

Of course, there were earlier networks. The Phoenicians' development of a written alphabet meant information could be both stored and moved. The Egyptians created a rudimentary postal system. The networks of roads the Romans built to connect the empire remain important pathways today. The activities of Western scriptoria painstakingly reproduced and preserved knowledge, while Eastern cultures aggressively explored its production at scale.

But the Mainz printing shop is a fulcrum of history.

What made Gutenberg's innovation so decisive were its multiple consequences. The original information network was the seed that permitted new ideas to germinate. Breaking information into its smallest unit structured the relationship between information and technology for subsequent centuries. And its redirection of information from sheltered silos to an outward surge recast the nature of inquiry, economics, and social structures.

Gutenberg himself was an example of how fifteenth-century society and economic activity were starting to flatten. The new middle class into which he had been born was squeezing into the social and economic space between the nobles and their serfs. Gutenberg's discovery continued that pattern with a similar flattening of the information hierarchy.

Modern economists describe this as a period of "interim technologies" that contributed to the explosion in the number of middle-class workshops.[54] Gutenberg's mechanization of the output of scribes was arguably the first such mechanical transformation of a handicraft.[55] Printing was also a precursor to the capital-intensive production models that followed, in which development, component inventory, and mechanical fabrication devoured capital before producing any return. "Making Bibles in Mainz did not differ fundamentally from making Fords in Detroit," it has been observed.[56]

But Gutenberg's press was not just turning out fenders; it was the platform on which discovery and innovation rested. It was a networked technology that begat a cumulative process of innovation that redirected the course of human events. For this reason, the narrative of how networks define who we are begins its modern journey with the network created by the movable-type printing press.

The vision of decomposing information to create it anew would become *the* foundation for subsequent information revolutions. We will see Gutenberg's disaggregation-reassembly idea reappear as the underpinning of the telegraph, then of the digital computer, then of the internet.

The Mainz workshop was also the beginning of a directional trend in the flow and use of information. Printing began a new pattern for the organization of information on a dispersed basis. Although rudimentary, the propagation and portability of printed information began the trajectory that manifests today in diverse information repositories connected by a pervasive network.

Gutenberg's discovery also demonstrates how connectivity alters the structures of authority. As the merchant printers' distributed authority began to destroy controlled-access information silos, the disaggregation of centralized authority over ideas and institutions followed.

Gutenberg humbles our self-centered assumption that today's information age is some kind of unique experience. In relative terms, Gutenberg's discovery occurred not that long ago. In the early part of the twenty-first century, we are closer to Gutenberg's success than the Mainz discovery was to the fifth-century beginning of the Middle Ages. Because of what the Gutenberg network launched, however, far greater revolutions have occurred in the past 550 years than in the preceding millennium. Gutenberg's final legacy was the acceleration of the flow of information, and thus the acceleration of change. The networks that followed the original information network further increased the velocity of information and thus the velocity of change.

Connections

Alfred Vail, the often overlooked assistant of Samuel F. B. Morse, stood before a printer's type box late in 1838 and linked Johannes Gutenberg to the world of electronic networks. Vail was in the printing shop to follow through on a Gutenberg-like inspiration about the most efficient means to transmit information electronically. While there is some historical debate as to whether Morse code was conceptualized by its namesake or actually by Vail, its core insight is indisputably Gutenberg's: information must be broken into its smallest part before it can successfully interface with technology.[57]

Later in this book we will explore the invention of the telegraph. That development would not have been one of the great transformational networks without its adoption of Gutenberg's insight. Early efforts with telegraphy had taken an approach similar to the early efforts at printing. Just as wooden plates were cut to contain all the information on a page, Samuel Morse's early concept of the telegraph was too broad to be practical. The original Morse design was an apparatus with a pencil attached to an electromagnet. As a paper tape moved under it, the pencil would make marks that spelled out numbers. Those numbers would then be referenced to a codebook in which each number corresponded to a specific word.

"I am up early & late, yet its progress is slow," Morse wrote about the "most tedious, never ending work" of compiling a dictionary that assigned a unique number to every word. In essence, Morse's early effort was to de-invent the alphabet and replace the words it spelled with unique numbers. The number 36, for instance, represented the word "abash," 37 its past tense "abashed," 38 "abashing," 39 "abate," and so on through the alphabet. The word "Wednesday," for instance, was coded as 4030.[58]

As in pre-Gutenberg printing, Morse was more focused on the final result—in his case, a message sent over distance—than on the productivity of the process to achieve that result. It was the same conceptual mistake made in early attempts at printing, which visualized information as a whole rather than as the sum of its parts.

The insight that the telegraph should transmit the components of a word rather than the word itself was a Gutenberg-like moment. That insight transformed the telegraph from a cumbersome curiosity to a highly productive information technology. It was for the purpose of designing this code that Alfred Vail visited a local printing shop and counted the number of pieces of type in the type box for each individual letter. The simplest signal (a single dot) was assigned to the most frequently used letter (*e*), with increasingly complex combinations attributable to other letters based on the volume of each letter in the printer's type box.

The exceedingly simple but powerful concept of breaking information into its smallest part so that it can be subsequently manipulated is the root concept in information technology. The introduction of that concept into Western civilization is the legacy of Johannes Gutenberg.

Gutenberg's breakthrough continues its connections through today's digital networks. The digital computing engine disassembles information into the smallest possible unit, the numbers zero and one. The fishnet-like digital network that is the internet does a Gutenberg-like disassembly of information into small packets and subsequently reassembles them at their destination, just as Gutenberg put disassembled pieces of type together to produce a new page.

The decade-long quest of Johannes Gutenberg that reached its conclusion in 1450 is the foundational discovery on which our knowledge and inspiration stand today. We connect with the man from Mainz because he made human inquiry a networked activity. One result of that networked activity, the scientific method of hypothesis and debate, touched off an explosion of discovery that has continued unabated. The form of that activity, breaking information into its component parts, thrives today as the interface between information and technology.

Gutenberg's discovery not only brought us to today, it is also at the heart of what drives us forward to tomorrow.

Three

The First High-Speed Network and the Death of Distance

The congressman arose in the hush of early morning darkness. Turning up the lamp, he gathered his effects. Abraham Lincoln had a 6:00 a.m. train to catch. He was heading home to Springfield, Illinois. The second session of the Thirteenth Congress had adjourned.

The trip Representative Lincoln was beginning was one to be endured rather than enjoyed. It would consume eleven days and three different modes of transportation. The journey was a powerful manifestation of how the forces of geography ruled the still young nation.

Early that morning, March 20, 1849, Lincoln boarded the Baltimore & Ohio Railroad for a daylong 178-mile ride to the end of the line at Cumberland, Maryland. There he transferred to a nine-passenger stagecoach for a rib-busting twenty-hour transit over the Appalachian Mountains to Wheeling on the Ohio River. Three days on a riverboat moved him 1,100 miles down the Ohio and then up the Mississippi River to St. Louis, where he alighted March 26. From there it was five more days aboard another bouncing stagecoach before he finally arrived home.[1] The completion of such an arduous trip was newsworthy. "Honorable A. Lincoln arrived in this city on Saturday evening," the *Illinois Journal* reported.[2]

In the mid-nineteenth century, the nation was a united group of states in name only. Geographic distance and natural encumbrances broke the country into semiautonomous isolated regions. Commerce and communication were an exercise in overcoming the impediments imposed by nature.

Yet the nation's future lay in the vastness of the land beyond the Eastern Seaboard. Some called it "Manifest Destiny." Whether those distributed parts of the nation could ever be linked would define that destiny.

Twelve years after the arduous trip home, the president-elect of the United States proceeded in the opposite direction. Starting on February 11, 1861, Abraham Lincoln's trip from Springfield to Washington was a triumphal tour of eleven of the nation's major cities. The entire 1,900-mile journey was completed by rail, much, if not most, of it over rails that had not existed at the time of his earlier journey. Had Lincoln chosen to go directly to the capital, the route that took him eleven days and cost much discomfort only a dozen years earlier could have been made entirely by rail in only two days.

With amazing speed, the railroad had begun to connect the nation. At the time of Lincoln's earlier trip there were approximately 8,000 miles of track in the United States; in February 1861 more than 30,000 miles of steel ribbon crisscrossed the nation.[3] By abolishing the absolute control of geography and the dictates of distance, those ever-extending steel ribbons created the interconnected reality we today take for granted.

As he gazed out the train car window on his journey back to the nation's capital, the stark contrast between his trips could not have been lost on Lincoln. The old world of isolated geographic sections of the country, each with its own customs, mores, and "peculiar institutions," was becoming interconnected. The question of whether the union would survive the changes that resulted from such interconnection would define the coming trials of the sixteenth president of the United States.

The Death of Distance

At the dawn of the American experiment, geography ruled the new nation. Statesmen of the day pondered whether the geographically vast but poorly interconnected nation could be held together.

The founding fathers' republican concept had never been attempted outside compact geographic areas such as Greek city-states or Swiss cantons. The question of whether a democratic republic could be held together in a large and poorly interconnected geography weighed heavily. When the First United States Congress convened in New York City in 1789, this fear was visibly clear. The time required to travel to the meeting from the thirteen states meant that it took weeks before there was a quorum to conduct business. From the relatively close city of Boston, for instance, a stagecoach traveling eighteen hours a day still took six days to reach New York.[4]

Two-thirds of the country's population was hemmed in by the Appalachian Mountains along little more than a fifty-mile-wide strip hugging the Atlantic Ocean. Coastal shipping moved passengers and freight up and down the Atlantic Seaboard. Inland, the waterways were the highways of commerce, but because they typically flowed north to south, these paths were not that helpful in efficiently uniting the nation's westward expansion from the eastern coast.

In 1810 the mayor of New York City, DeWitt Clinton, soon to be the governor of the state, proposed that man should do what nature had not: construct an east-west waterway. Clinton's state had two major advantages that made this possible. First, the river that Henry Hudson had thought might be the Northwest Passage and that bears his name stretched from the Port of New York 315 miles inland to Albany, providing the foundation for the westward push. Even more important, however, was that at Albany, the Mohawk Valley provided one of the only breaks in the 2,500-mile Appalachian chain. Through that gap Clinton proposed digging a canal to link Albany with Buffalo on Lake Erie and from there to the nation's interior via the Great Lakes.

The 364-mile Erie Canal took eight years and $8 million to build before it began operation in 1825. A human-engineered surrogate for nature's pathways, the canal was the greatest engineering project the continent had ever seen. Using picks and shovels, gangs of men dug a forty-foot-wide, four-foot-deep ditch through the wilderness. Eighty-three locks were constructed to handle the gradient of the land.

The resulting pathway reduced the cost of transporting bulk cargo from 20–30 cents per ton-mile to 2–3 cents. It cut the Buffalo–New York trip from a month of hard travel over bad roads to only ten days.[5] The result supercharged the Port of New York, making it the nation's most active and building the city into the country's largest.

It is not hard to imagine the reaction of the residents of the other Atlantic coast ports as they watched New York's canal-driven success. Prior to the canal, Philadelphia had been a busier port and thus a bigger city. Likewise, the competitive port cities of Baltimore, Boston, and Charleston all feared the impact of the New York canal and sought solutions to their geographically impaired access to the West.

The obvious answer was to copy New York's ditch. The problem for the other cities, however, was the absence of a similar low-level passage through the mountains. The Pennsylvania legislature, in an effort to return the Port of Philadelphia to supremacy, funded the Main Line of Public Works, almost 350 miles of canals and inclines. When the Main Line confronted an obstacle, the goods were off-loaded onto inclined planes, dragged over the obstruction, and reloaded onto another barge on the other side. The enormous cost to build and operate such a process, however, assured it would neither be profitable nor successfully replace New York's canal.[6]

In the same year that Governor Clinton's canal was completed, however, the first scheduled steam railroad began operation in England. It didn't take long for the technology to leap the Atlantic.

Two separate events on July 4, 1828, three years after the completion of the Erie Canal, symbolized the transformational change under way. In Washington, D.C., President John Quincy Adams broke ground for the Chesapeake & Ohio (C&O) Canal. In Baltimore, forty miles to the north,

Charles Carroll, the sole surviving signer of the Declaration of Independence, used a silver spade to turn the ground and launch construction of the Baltimore & Ohio (B&O) Railroad.

Both the canal and the railroad were designed to do for their city's commerce what the Erie Canal had done for New York. The ways in which they sought to accomplish this goal, however, were from two different eras. The C&O Canal was a continuation of nature's network. The railroad, in stark contrast, was mankind asserting dominion over nature.

While Baltimore's rail line first stumbled with multiple approaches to the new idea—trying horse-drawn railcars, cars propelled by horses on treadmills, and even sail power (Baltimore was, after all, a sailing town)—the B&O ultimately embraced the new English invention of steam locomotion. In 1830 the B&O became the first American railroad to haul both freight and passengers by steam on a regular schedule.

Laying tracks atop the land was faster and more practical than excavating a canal. By the time the C&O Canal had reached its terminus at Cumberland, Maryland, in 1850, the B&O Railroad had been there for eight years. Not only could railroads be constructed more quickly, they also moved their contents infinitely faster. Whereas canal boats crept along at around four miles per hour, the early locomotives sped over the rails at blazing speeds of up to twenty miles per hour (and they kept getting faster).

Moving people and products faster meant moving them farther. It was the perfect marriage—the seemingly endless land mass of the United States and the virtually boundless energy of the steam locomotive. The result was the beginning of national integration. The miles that had previously segregated populations melted away. The railroad became the domestic melting pot for both individuals and commerce.

Thanks to the railroad, American agriculture changed from a subsistence or local market activity to a commercial enterprise. Farm produce's low value relative to its high bulk had always limited its transportation over any distance. The cost of carting a sixty-pound bushel of wheat constrained consumption to within close proximity of where it was grown.[7] The railroad changed that equation by making it possible to haul bulky products to a distant market at a rational cost. One observer described railroads as a centrifugal pump drawing harvests further from their fields.[8]

The railroad did the same for the bulky raw materials necessary for manufacturing. The United States was late to the Industrial Revolution because it lacked the necessary power sources (other than the water power of the Atlantic fall line). The railroad solved that problem by inexpensively hauling coal from the mines to fire the furnaces and fabricate other rail-delivered natural resources into products. Then the railroad distributed the finished products to an interconnected national market.

The printing networks had sped the transformation of Western thought. Now the rail networks were rapidly transforming commerce and the patterns of everyday life. Every town the railroad touched was forever altered by it, and every town it bypassed was disadvantaged. Nowhere is the power of such network connectivity more evident than in the story of how the railroad determined the divergent courses of two western cities: St. Louis, the Gateway to the West, and the small Illinois village the Indians called "Chicagou," meaning "the wild garlic place."[9]

On October 25, 1848, the inauguration of the first railroad into and out of Chicago set off a string of events that "would change not merely the nation, but the world."[10] On that day the Galena & Chicago Union Railroad celebrated construction of its first eight miles of track at the Oak Ridge (now Oak Park) end of the line. As the name suggested, the plan was to link Galena—in northwestern Illinois on the Mississippi-feeding Galena River—with Chicago on Lake Michigan. The role of the railroad, it was assumed, was to facilitate the movement of products to and from those waterways. What occurred, however, was the beginning of a process in which the new network replaced the old.

The directors of the Galena & Chicago Union Railroad and other VIPs rode behind the steam locomotive *The Pioneer* to the inaugural celebration in Oak Ridge. After appropriate remarks and other festivities, the company directors prepared to reboard their coach for the return to the city. One of the directors, however, had noticed a farmer in the assembled group of onlookers sitting atop an ox-drawn wagon piled with sacks of wheat and hides. When asked, the farmer indicated he was headed to Chicago to sell his products. The director purchased the man's produce and the train hauled it back to Chicago.[11] The cargo was the first revenue for the railroad. Far more significant, it was the beginning of a pattern of economic activity. Within a week there were thirty carloads of wheat at the end of the line awaiting shipment to Chicago.[12] The railroad was profitable from its first day of operation.[13]

The *Chicago Daily Journal* observed of the opening of the Galena & Chicago line, "The 'Iron Horse' is now fairly harnessed in the Prairie Land, and the freedom with which he travels, betokens his satisfaction with the bounteous and almost unlimited pasture field."[14] By expanding into that "bounteous and almost unlimited pasture," the railroad opened up markets for the products of that land and transformed agriculture from an activity of self-sufficient survival, or at best one of local markets, into a far-reaching commercial enterprise.

Soon forests were felled and sod busted for railroad-induced agricultural production. Chicago became the great hub that linked the products of the plains with the tables of eastern consumers. Within four years of the run from Oak Ridge to Chicago there would be more grain reaching Chicago over steel rails than by wagon and canal combined.[15]

The flow of commerce, which had previously been forced into the north-to-south path of the region's waterways, changed direction to west-to-east. Previously, if the bulky product of western farms was to move eastward, it had to move down the Mississippi to New Orleans for subsequent ocean shipment to eastern markets. As railroad networks expanded, however, the products of the bounteous pasture moved to Chicago for onward shipment due east.

While the railroad created commercial centers such as Chicago, Indianapolis, Columbus, Toledo, and Detroit, it also shaped small towns. Feeder lines reached into the hinterland like ganglia spreading out of the network hubs.[16] Whereas waterways commanded commercial activity to come to them, railroads went anywhere, including directly to the fields and mines that produced the bounty of the western lands. Many of the heartland towns of Ohio, Indiana, and Illinois owe their existence to the arrival of the railroad.[17] Similarly, railroads diminished the role of the cities built along nature's networks. Nowhere was this truer than for the Gateway to the West, St. Louis.

Located on the western side of the Mississippi River, St. Louis was the center of waterborne commerce along the Father of Waters. At the time of the Galena railroad inauguration, Chicago was a small town with a few thousand inhabitants, while St. Louis was the principal city of the West.

Unfortunately for St. Louis, however, the "Big Muddy" stood between the city's perch on the river's western bank and the expanding rail lines connecting to eastern markets. A city controlled by watermen saw nothing illogical with bringing a load of cargo from the west, placing it on barges to the eastern shore, and then putting it on railcars. The city fathers of St. Louis refused to even consider a bridge to connect the city to the lines from the east.[18]

Chicago's leaders, on the other hand, aggressively pushed for rail connections, even flouting the law to assure their connection to the east. When in 1851 it appeared as though the rails from the east might pass south of Chicago, the locals swung into action.[19] Mass protests were organized. The Common Council (city council) appropriated the sizable sum of $10,000 to fight the plan. The mayor and U.S. senator Stephen A. Douglas journeyed to the financial markets of New York to discourage funding of the bypass project.[20]

The matter was ultimately resolved when one group simply started laying down track without the necessary permission of the state. For six and a half miles, with questionable legal authority, the tracks crossed the countryside to meet the rails from the east and bring them to Chicago.[21] A look at the railroad map of the time shows a curious dogleg northward to Chicago rather than a straight shot that could have connected the line from the east with the southern end of the canal leading to Chicago and Lake Michigan before proceeding straight on to the west.

Chicago's vision was transformative. By 1854, six years after the first run of the Galena & Chicago Union Railway, Chicago was the railroad center of the west. By 1860, there were fifteen railroads converging on the city with a hundred trains coming and going daily.[22]

Chicago, not St. Louis, had become America's junction. "St. Louis was soon 'eating Chicago's dust' as a grain entrepôt," one railroad historian colorfully explained, "and would spend the next fifty years attributing its neglect to almost anything save the truth, which was its backwardness in providing handling facilities and a railroad bridge across the Mississippi to make a through rail route to the East."[23]

By ending geography's stranglehold on the affairs of civilization, the steam locomotive had begun a new network revolution. There could be no more graphic example of how speeding locomotives were accelerating change than the tale of how, in the space of a decade, the rails opened the prairie and made the sleepy, swampy village on the banks of Lake Michigan the nation's Second City.

Old Concepts, New Results

Like the printing press and every one of the network technologies that would follow, the steam railroad was the result of a convergence of preexisting technologies. In this instance, it was the flanged wheel and the steam engine.

Evidence suggests that as far back as biblical times, wagons rolled on tracks to carry cargo.[24] By the seventeenth century, probably beginning in the coal mines of the Ruhr valley, flanges were added to one side of the wheel to help hold it on the rail.[25] By the mid-eighteenth century iron rails were replacing wooden ones. Not only did the iron rails have increased durability, but since a metal wheel rolling on a metal rail has the lowest coefficient of friction of any form of carriage, the new rails allowed an animal to pull a greater load. Throughout the coal-producing regions of Europe, horse-pulled rail wagonways proliferated to bring heavy loads out of the mines as well as to haul the mined product to rivers and canals.

The challenge at these coal mines was that the deeper they went, the more groundwater seeped into the shafts. Often miners would stand knee-deep as they hacked at the coalface. It was therefore a breakthrough when in 1698 the Englishman Thomas Savery developed an "engine to raise water by fire." Savery's pump was based on the physical reality that steam occupies 1,600 times the volume of the water from which it is produced.[26] Thus a small amount of water could generate a tremendous volume of steam, and cooling that steam back to water created a vacuum. Savery conceived the pump to use that vacuum to suck water from the mines. Because the suction dissipated with distance, however, Savery's solution was effective only to a depth of sixty or seventy feet.[27]

Thomas Newcomen solved the suction problem in 1712 by harnessing another set of established concepts. Like the "engine by fire," Newcomen's engine created a vacuum by condensing steam. Instead of using the vacuum to suck water, however, Newcomen used it to drive a piston.

The revolution in steam was yet another residual effect of Johannes Gutenberg's discovery. Like almost all of those who developed new techniques to harness steam, Thomas Newcomen had little formal education, but he had access to books on vacuums and pistons, from which he taught himself. Based on that reading, Newcomen knew that turning water to steam within a closed container creates pressure and that cooling that steam back into water produces the opposite effect, a vacuum. He harnessed that action to a piston within a cylindrical chamber with an open top. The production of steam pushed the piston up the cylinder; then, when the steam condensed and created a vacuum, the atmospheric pressure brought it back down. To this piston Newcomen attached a seesaw-like beam that connected to reciprocating-force pumps down in the mine. As the steam pushed the piston upward, the pump handle descended; then, as the atmospheric pressure drove the piston downward, the seesaw would draw the pump lever upward.

Newcomen's engine, while a boon to coal mining, was inherently inefficient. Since the piston cylinder was also the condenser, it had to be reheated after each cooling. As the cold cylinder needed to climb back to at least 212 degrees Fahrenheit to produce the necessary steam to push the piston back up, the engine suffered from long cycle times.[28] As a result of this inefficiency, the Newcomen engine's fuel consumption was considerable. Fortunately, the engine's principal use was at the mouth of the mine that produced the fuel it required.

Half a century passed before the spring of 1765, when James Watt solved the Newcomen inefficiency and became the "father of steam." Watt's idea was to separate the heating and cooling functions, enclose the piston cylinder, and reverse the principal force acting on the piston. In Newcomen's engine the power came from the pressure of the atmosphere forcing the piston back down the cylinder. The power from Watt's steam engine came in the other direction: the force of the steam being introduced into the chamber to drive the piston upward.

To accomplish this, Watt bled the steam into a separate piston-driving chamber surrounded by cool water, where the steam would rapidly condense. The hot part of the engine stayed hot to continually produce the steam to drive the piston, while in the chamber with the piston that steam rapidly condensed, to allow another blast of steam to drive the piston again. This process was more powerful and more reliable than the Newcomen engine, and as the approach did not necessitate a continual heat-cool-reheat-recool process, the Watt steam engine used only about a third of the fuel its predecessor used.

While Thomas Newcomen harnessed steam, and James Watt made it the engine of a new era, two realities constrained the future of the steam engine.

The first constraint was the inability to increase its power. Increasing the pressure within the engine in order to increase the power of the piston was deemed suicidal. The father of steam himself argued vociferously against the safety of a high-pressure engine.[29]

The second constraint on steam engines was their immobility. At several tons, they were simply too big and heavy to transport. The power the engine produced was insufficient to overcome the friction of a rolling wheel.

One man tackled both constraints. In 1800, a Cornish inventor named

Richard Trevithick accomplished what James Watt said only a looney would try. At the Wheel Hope copper mine in western Cornwall, Trevithick constructed an engine of "strong steam" in which greater pressure produced greater power. Trevithick's success prompted the owner of a Welsh ironworks to lure the inventor from Cornwall to Wales.

Then came the wager.

Sometime in 1803, outside the Welsh town of Fetidsty (a name that conjures the town's reported ten-pigs-to-one-person ratio), the owner of the ironworks stood with a group of friends and watched a horse pull wagons along a tramway. They debated whether the horse could ever be replaced with steam power. The boss saw an opportunity. He bet 500 guineas that his new employee, Trevithick, could build a steam engine capable of hauling ten tons of iron ore the nine and a half miles of the tramway. It was no idle wager: at the time, the average worker earned fifty guineas per year.[30]

On February 21, 1804, a "strong steam" engine built by Richard Trevithick did what had never been done before. "Yesterday we proceeded on our journey with the engine," Trevithick wrote, "and we carried ten tons of iron in five wagons, and seventy men riding on them the whole of the journey . . . the engine, while working, went nearly five miles an hour."[31]

Unfortunately, the inventor's "while working" aside spoke to the trial's ignominious end. On the return journey a bolt broke in the engine and the boiler leaked. The fire was dropped, and the now steamless engine was drawn by horses the last couple of miles back to the starting point. The impact this had on the wager is unknown. The results of the bet, however, were monumental. A self-propelled engine, riding on rails, had pulled a collection of connected cars containing both people and product.

By pulling ten tons over a track for multiple miles, Richard Trevithick had proved that the tremendous weight of a steam engine could be reduced to a manageable amount while still generating power sufficient to draw a heavy load. It was a profound event rooted in a simple concept: increasing the heat of the boiler.[32]

Trevithick's breakthrough was to eliminate altogether Watt's breakthrough, the condenser. He built his strong steam engines to allow the steam to escape directly into the outside air rather than into a condenser. Instead of cooling the steam to produce a vacuum, he simply purged it, to the same effect. Observers of the escaping steam nicknamed his engines "puffers."

Allowing the steam to escape meant a loss of pressure in the cylinder.[33] The challenge became to increase the pressure to replace—and exceed—the loss. Trevithick's solution was to increase the heat. "My predecessors put their boiler in the fire," Trevithick explained. "I have put my fire in the boiler."[34]

Newcomen, Watt, and others had envisioned the steam-engine boiler as a giant teakettle sitting atop a source of heat. Some of the heat energy was captured by the boiler, but a huge amount simply escaped around it. By moving to an internal boiler surrounded by water Trevithick captured more heat energy and turned it into steam. It was estimated that for every thirty degrees of increased heat, the steam power would double.

It was a huge breakthrough: using fuel to produce an incremental increase in heat resulted in an exponential increase in power. And the elimination of the condenser provided the additional benefit of eliminating weight.[35] Trevithick's final breakthrough was the smokestack. By forcing the escaping steam into a chimney, he created a vacuum behind it that pulled more air into the fire, giving it more oxygen, making it burn even hotter, thus feeding his heat-to-power conversion.

The Trevithick engine, after so brilliantly demonstrating strong steam and the commercial application of a self-propulsion apparatus, suffered an ignoble fate. It was grounded, its wheels were removed, and it was reduced to powering a winch that pulled a cable attached to the ore cars.[36] Though the steam engine had worked, the rails on which the engine traveled had not. To ensure the horses that typically hauled the ore carts would not stumble, there were no crossties between the rails. Instead, each rail was anchored atop a block that was independent of the block anchoring the opposite rail. As the five and a half tons of the Trevithick engine rolled down the tramway, the lateral force cracked the rails from their moorings. The heroic achievement of steam locomotion was inhibited by the limitations of the rails on which it rode.

Steam may have been harnessed to ride on rails, but the manner in which the rails were constructed slowed its application for more than a decade. Finally, in 1816, another self-taught engineer, George Stephenson, solved the problem by patenting the idea of crossties that connected one track to the other.[37] The outward pressure of weight on the rails would be constrained by tying the rails crosswise with each other.

Shortly thereafter, Stephenson, who had also been building strong steam engines, was hired by a group that wanted to build a public tramway over the twenty-five miles between the towns of Stockton and Darlington in northeastern England.[38]

On September 27, 1825, Stephenson's creation, the Stockton and Darlington Railway, began operations. The new railroad moved several hundred people twelve miles in cars that resembled coal wagons with benches. It was the first commercial undertaking to harness the speed and endurance of the steam locomotive to carry both goods and passengers. "The adaptation of rail-ways to speed was never, we believe, thought of till the opening . . . of the celebrated Stockton & Darlington rail-road," an English newspaper observed in retrospect five years later.[39] That speed was provided by Stephenson's steam engine *Locomotion No. 1*, from which the term "locomotive" is derived.

Five years later, another Stephenson creation linked the Port of Liverpool, the busiest in the world, with the cotton capital of the world, Manchester, thirty-one miles away.[40] The Liverpool and Manchester Railway was an instant financial success. Beyond its initial application, the line was a harbinger of things to come. Built principally to haul cotton and coal, it soon became a magnet for other products. Families' dinner plates in towns along its path, for instance, began to include the perishable foods that had once been too bulky to transport and thus too expensive for the typical family to buy. Newspapers began to circulate freely between the two cities.[41]

The second great network transformation was under way. Steam rolling on steel recast commerce and redefined the patterns of life.[42] It was, as one commenter observed, "The completest change in human experience since the nomadic tribes became rooted in one spot to grow grain and raise cattle."[43]

Transformation's Tumult

Such a change in the human experience was bound to bring with it turbulence and scrutiny. In 1829, the year before the Liverpool and Manchester Railway opened, steam locomotion leapt the Atlantic. Its result was to shape—not just connect—geographies. Unlike Europe, with its older, well-established cities, the railroad's path across the American emptiness would turn crossroads into cities, convert empty prairies into productive farms, and usher in a new industrial era—all to varying degrees of consternation.

Much as with the British experience, the American railroad began on a horse tramway. A northeastern Pennsylvania canal company had built a tramway between its ditch and some anthracite mines nine miles away. Hoping that the new technology of steam locomotion might help overcome the tramway's steep grades, the chief engineer sent an assistant to England to observe what George Stephenson was developing. Two locomotives were purchased for shipment across the Atlantic.

The *Stourbridge Lion* (named after the English town where it was constructed) pulled the first load over the tramway on August 8, 1829. Unfortunately, it suffered the same fate as Trevithick's locomotive. The rails could not support the engine's weight. It went into a storage shed and never emerged.[44]

A year later, the B&O Railroad, which had tried everything from horses to sails as a means of propulsion, finally tried steam locomotion. On August 28, 1830, a small steam engine named for P. T. Barnum's diminutive star Tom Thumb hauled thirty-six people at speeds of up to eighteen miles an hour.[45] The engine's builder, Peter Cooper, had learned from the *Stourbridge Lion*'s failure and built an engine light enough for the rails over which it would travel.[46]

Tom Thumb introduced speed to North America. Racing at four and a half times the speed of any other conveyance, *Tom Thumb* was both a marvel and a mystery. The train's owners and occupants first questioned whether the human body could endure such speed. Many of the passengers on *Tom Thumb*'s first run were human guinea pigs who brought along paper and pencil to test whether cogent thought was possible at such speed.[47]

In a foretelling of challenges to come, the incumbent stagecoach operators tried to nip the puffing competitor in the bud, arranging a race between a horse-drawn carriage and the locomotive. The horse shot out of the blocks as the engine built up steam. Once fully under way, however, *Tom Thumb* closed the gap and surged ahead. The puffer held a comfortable lead until a blower belt slipped and the horse galloped past. The locomotive had lost the race, but not the contest. The leadership of the B&O redirected its efforts into becoming an all-steam railroad.[48]

Charleston was the other Atlantic city racing to railways as an alternative to New York's canal. On Christmas Day 1830, behind the American-built *Best Friend of Charleston,* a chain of cars carried dignitaries for six miles over the new Charleston and Hamburg Railroad.[49] As in Baltimore, the experience of speed was surreal. The *Charleston Courier* wrote of the event, "The one hundred and forty-one persons flew on the wings of the wind at the speed of fifteen to twenty-five miles per hour, annihilating time and space . . . leaving the world behind."[50]

For a civilization whose land speed had forever been limited to the pace of an animal, the concept of a speeding locomotive was often too much to grasp. "What can be more palpably absurd and ridiculous than the prospect held out of locomotives travelling twice as fast as stagecoaches!" Great Britain's *Quarterly Review* exclaimed about the Liverpool and Manchester Railway. "We trust that Parliament will, in all railways it may sanction, limit the speed to eight or nine miles an hour."[51]

Indeed, members of Parliament were skeptical of speed. When John Stephenson, son of the designer of the Liverpool and Manchester line, appeared before the legislature seeking a charter for a new railroad, the following colloquy took place with one MP:

"Well, Mr. Stephenson, perhaps you could go 17 miles an hour?"

"Yes."

"Perhaps 20 miles an hour?"

"Certainly."

"Twenty-five, I dare say. You do not think that impossible?"

"Not at all impossible."

"Dangerous though?"

"Certainly not."

"Now, tell me, Mr. Stephenson, will you say that you can go 30 miles an hour?"

"Certainly."

At which, it is reported, "they all leaned back in their chairs and roared with laughter."[52]

On the American side of the ocean the speed of the steam locomotive was increasingly inescapable; the locomotive was an instrument of constantly increasing velocity. From the *Tom Thumb*'s eighteen miles per hour in 1830, the speed of steam locomotives steadily accelerated. Each increase in speed set a new record, until by 1893 the New York Central's *Empire Express* was the fastest moving thing on Earth at 112.5 miles per hour.[53]

And it wasn't just the locomotives that were speeding along. With great speed a nation defined by great distances embraced the network. It was the death of distance.[54] Just ten years after *Tom Thumb* and the *Best Friend of Charleston,* the total rail mileage in the United States exceeded that of canals.[55] By the mid-1850s, the nation with less than 5 percent of the world's population had track mileage that nearly equaled the combined mileage of the rest of the world. By 1860, only thirty years after *Tom Thumb,* there were 30,000 miles of railroad track in the United States.[56]

As speed recast everyday life, the change was not always welcome. The first recorded use of the phrase "Oh, for the good old days" was a lament about steam locomotion. Writing in 1844, Philip Hone, the former mayor of New York, complained, "Railroads, steamers, packets, race against time and beat it hollow. . . . Oh for the good old days of heavy post coaches and speed at the rate of six miles an hour!"[57] As we have previously seen, one journalist warned that the speed of steam locomotives "will give an unnatural impetus to society, destroy all the relations that exist between man and man, overthrow all mercantile regulation, and create, at the peril of life, all sorts of confusion and distress."[58] He was right.

Like the Renaissance, which would have remained a local phenomenon if not for the printing press, the American Industrial Revolution would have continued to lag behind European advances—maintaining the United States as a "lesser-developed country"—without the widespread adoption of the railroad.

The United States was late to the Industrial Revolution for two principal reasons: the lack of adequate power and the absence of an adequate market.[59] The fall line that ran along the East Coast provided power from cascading water but geographically constrained early industrialization. The poor connections among cities and regions created a balkanized marketplace that further constrained both the stimulation and the sustaining of scope and scale production. The spread of railroads created a synergy that vanquished those restraints.

By bringing coal to fire stationary steam engines, the railroad liberated industrial production from the Atlantic fall line. Then, by distributing the finished product to an ever-expanding and increasingly interconnected market, the railroad both created and serviced demand for industrial output.

The railroad was an industrial perpetual-motion machine. Created at first to haul the products of mines, railroads soon became the largest consumer of one of those mines' extracts: coal. In a similar manner, railroad construction drove demand for the high-volume production of other manufactured goods. Steam-powered mills were required to produce the lumber used for ties and trestles. Likewise, iron and steel production expanded to meet the demand for engines and rails. These and other products were, of course, delivered by rail.

Beyond the demand for raw materials, railroads required an unprecedented industrial support apparatus to keep the trains rolling. The maintenance of engines and rolling stock, and the production of the necessary industrial equipment for the task, fed the growth of mechanical industries and the spread of mechanical skills. By 1870, railroads were consuming 20 percent of all the machinery produced in the United States and 40 percent of all the rolled steel.[60]

Railway demand for industrial production created a slipstream behind which other industries could grow by taking advantage of the railroad's market momentum. The tooling up of various production activities to supply the railroads created techniques and capacity that could be applied elsewhere. Machine tools necessary for producing and servicing engines and rolling stock were adapted to the production of new equipment for farm and factory. Strong, low-cost girders like those that spanned rivers soon rose upward into ever-taller buildings. The production of lumber and steel mills or the fabrication of industrial equipment for the railroads created a long tail of baseline capabilities on which other commercial activities could build. And the use of the railroad itself was available at low marginal cost after the high fixed costs of infrastructure development, as its economic base had been addressed by hauling extracted commodities. Both raw and finished products, including newspapers and the mail, began to hop a ride on the train whose pathway and profit had been directed by high-bulk cargo.[61]

The railroads were the first business to operate at scale. At a time when almost every business was a family affair run out of one place, the railroad was massive and complex. As it became the nation's largest employer, the demands of managing railroad operations birthed bureaucracy. The construction of rail lines alone was the greatest management challenge ever undertaken. The delivery of large caches of supplies and materials across great distances had to be coordinated with the activity of tens of thousands of workers who likewise were spread across the country.

Once the railroads were constructed, operations over the rails created new managerial challenges to keep the trains staffed, serviced, supplied, loaded, and running on schedule. Not surprisingly, the management ranks of the railroads were filled by men with experience in the nation's only other large institution, the U.S. Army. Approximately 120 graduates of West Point became senior railroad executives while the subordinate ranks were filled by countless others who had previously served under them.[62]

Such a large undertaking required systemization and discipline. Managerial concepts common today had their origin in railroading. In 1846, the B&O Railroad issued its *Proposed New System of Management*, which established functional chains of command. The Erie Railroad went so far as to define the responsibilities and authority of each of its 4,000 employees, along with establishing both standards of efficiency and common disciplinary procedures.[63] The result of these and other efforts created a common process with a hierarchical structure that, not surprisingly, resembled that of the military.

The railroad's new management model was soon embraced by other businesses. As the new scope and scale of industrial production for national markets harnessed masses of workers to a common process, corporate bureaucracy provided supervision and structure. This hierarchical organization also created a new path for individual development. Back on the farm there was little chance to move up the ladder, but in industrial institutions the promise of advancement into the middle class beckoned.

As railroads brought nature's bounty to central points for fabrication, they also transported the masses of workers necessary to operate the factories and provide support. At the start of the railroad era in 1830, there were ninety towns in America with a population of 2,000 or more. By 1860 the number had grown to 392.[64] The draw of these new urban hubs of economic activity remade the national map. Railroads created new urban centers, transporting workers from the fields and immigrants from the docks to the new urban America. By the time of the 1890 census, more Americans were living in cities than in rural areas.[65]

The growth patterns of the railroad and urban centers were duplicates of each other. And as crowds congregated in growing cities, individuals surrendered a portion of their independence. Urban life imposed a new social order built around corporate collectivism. Wide-open fields yielded to tenements and individual interdependence, spawning a host of new societal problems. Masses of people produced prodigious amounts of waste, which necessitated the development of public sanitation practices and disposal systems. Safe freshwater systems were required to keep the scourge of yellow fever from ripping through a city. As urbanites experienced illness en masse, the town physician was replaced by institutional hospitals that applied mass-production-like solutions to health-care delivery. And the need for an educated workforce saw the one-room school replaced by public education factories in which youthful raw material went in one end and emerged several years later sufficiently educated to become cogs in the industrial machine.

Increasingly, the technology that had successfully asserted its superiority over nature began to assert new authority over the affairs of man. Many of the things we today consider as bedrock traditions were, in their time, railroad-induced upheavals.

Railroading changed the law. It began with the sacred right of property ownership. English common law had recognized the authority of the state to take private land for public good (eminent domain). As the growth of the railroad required ever-increasing amounts of land across a specific trajectory, the railroad companies assumed this confiscatory right as inherent in their state-granted operating charter. Starched-shirt lawyers and accountants from faceless corporations descended to confiscate land that had been cleared and cultivated by the strength of the owner's back and the sweat of his brow.

The new network also brought with it a new potential for harm to both individuals and property. Again, the law changed to benefit the new technology. The legal concept of strict liability (that is, if you caused it you were responsible for recompense) began to erode. A new legal paradigm based on negligence took its place. No longer did determination of guilt revolve around who caused the problem; the circumstances of the event and whether the defendant had taken appropriate precautions determined culpability. Thus, if hot cinders from a passing locomotive set fire to your barn, the legal issue wasn't the source of the fire but whether the railroad had attempted to mitigate the flying cinders, such as by placing a screen across the opening of the smokestack.

Bending the laws of man to the will of the railroad was one thing. Soon, however, the railroads were bending the "laws of God." Even the sacred Sabbath succumbed to steam locomotion. Railroads originally refused to operate on the Sabbath, but the profit potential of moving people and products on Sunday soon began to erode such piety.

Exhortations from the pulpit and mass demonstrations to stop Sunday trains proved insufficient to halt the sacrilege. Rebuffed in their civil actions, the God-fearing turned to Lady Justice. Once again, however, the courts sided with the new over the traditional. A decision of the Supreme Court of Pennsylvania illustrates the emotional and political complexity of the issue: by twisting logic into a pretzel, the court authorized Sunday rail operation as an essential means for the poor to get to church.[66]

The railroads even encroached on what many considered a force of God: the determination of the time of day. In mid-nineteenth-century America there were approximately 100 local "sun time" variations across the country. Time of day was typically determined by a mutually acknowledged reading of the position of the sun against a local landmark (often the church steeple), to which local clocks were set. Illinois had twenty-seven different time settings, Indiana twenty-three, and Wisconsin a whopping thirty-eight. When travel ran four miles an hour, the time required to cover any distance mitigated local differences. As rails vanquished distance, however, the resulting mayhem was "no way to run a railroad." The Buffalo train station, for instance, had three different clocks, each giving a different time based on destination variations.[67]

Finally, in 1872, the superintendents of the various railroads convened in a General Time Convention to try to find a solution. The issue was so heated it took eleven years before even its proponents could reach agreement. On November 18, 1883, Railroad Standard Time went into effect, dividing the continent into five time zones (including one for Canada). The railroads, however, had no ability to impose the measurement on others.

In New York City and Boston, local time became Railroad Time. That counted for little in other parts of the country, however. In Columbus, Ohio, and Fort Wayne, Indiana, the time convention was ignored as a burden on the honest worker. The Indianapolis *Sentinel* editorialized, "The sun is no longer boss of the job. . . . It is a revolt, a rebellion. . . . People will have to marry by railroad time and die by railroad time. . . . We presume the sun, moon, and stars will make an attempt to ignore the orders of the Railroad Convention, but they, too, will have to give in at last."[68]

The federal government concurred with the rebellion against standard time. The attorney general of the United States forbade government offices from changing their time unless Congress passed a law pursuant to the convention. The people's representatives, fully aware of the political dynamite they were handling, took thirty-five years before finally enacting legislation on the issue in 1918.

Steam locomotives, however, waited for no one. Railroad schedules began to influence the timing of local activities. Instead of being determined by the pace of local life, the schedules of towns along the rail line were set by a remote power. Many of a community's activities were paced by the comings and goings of the town's most important network connection—a schedule determined not by local preference but by what was required to arrive at a time convenient for a distant destination.

That schedule and the new importance of timeliness created an opportunity for Richard W. Sears, the twenty-two-year-old North Rosewood, Minnesota, station agent for the Minneapolis and St. Louis Railroad. When in 1886 a local jewelry store refused a consignment of watches, Sears stepped in, agreeing to pay the watchmaker $12.00 for every watch he sold. Sears correctly recognized that the railroad's schedule made his passengers especially sensitive to timeliness, thus making them an ideal target audience for the products. For his sales force, Sears turned to his fellow station agents, who were already interconnected by a means of distribution. Within six months, Sears had netted $5,000, quit his job, and started the R. W. Sears Watch Company. The following year he hired his first employee, a watch repairman named Alvah Curtis Roebuck.

The rails, which had first opened the plains, and then transported salesmen and peddlers to the new market, also enabled the postal system's expansion. This expansion in turn made possible the placing of orders and the delivery of goods—the mail-order catalog business. Sears and Roebuck, along with other pioneers like Montgomery Ward, were only too happy to seize upon the new opportunity. Chicago, the rail hub of the nation, became the center of the mail-order business.

Pushback

Not surprisingly, as the railroad steamed over tradition, it was met by a countervailing force. Those opposed to the change warned that the racing passage of a steam locomotive would cause cows to stop grazing and hens to cease laying. Birds would drop from the sky from the smokestack's exhaust of poisoned air. And horses, replaced by engines, would become extinct, with the result that farmers of hay and oats would go bankrupt.[69]

"All conceptions will be exaggerated by the magnificent notions of distance.—Only a hundred miles off!," warned the Vincennes, Indiana, newspaper in 1830. "Why, you will not be able to keep an apprentice boy at his work! Every Saturday evening he must have a trip to Ohio to spend a Sunday with his sweetheart. Grave plodding citizens will be flying about like comets. All local attachments will be at an end."[70] An Ohio school board condemned railroads as "a device of Satan to lead immortal souls to hell."[71]

Stirring this pot of confusion and concern were those whose business would be negatively affected by the new network. In Great Britain, surveyors scouting pathways for rail lines were attacked by thugs hired by local businesses.[72] In the United States, local vigilantes, organized by those whose business would be affected by the railroad, emerged under the cover of darkness to tear up tracks laid that day.

The opposition didn't always resort to thuggery, however. In a precursor of the battles occasioned by subsequent network revolutions, public relations campaigns and legal and political challenges were mounted. "Every ploy known to shrewd local lawyers was used to keep things nice and cozy for local carting companies, freight forwarders, hack drivers, hotel and restaurant owners, local wholesale merchants, and anyone else who found a foothold in an environment of what might be called 'enlightened backwardness,' " observed railroad historian Albro Martin.[73]

As Philadelphia considered connecting two hitherto separate rail lines in 1839, for instance, a poster went up throughout the city. The sign said nothing about its sponsors, or their desire to protect their businesses of hauling people and freight between the two disconnected lines and selling food and sundries to those in transit. Instead, the placard portrayed the linkage as a dire threat to public safety—especially to women and children. The dominant image of the poster portrayed a carriage being upended by a racing locomotive as ladies and children scurried to safety. "MOTHERS LOOK OUT FOR YOUR CHILDREN!" the poster blared. As if the threat to physical safety weren't bad enough, in slightly larger type was an appeal to local pride with the warning that the connection would turn Philadelphia into a "SUBURB OF NEW YORK!!"

One of the classic efforts to stop the expansion of railroads came from the steamboat companies that had previously dominated western transportation. Because the superiority of the rails ended abruptly at the river's

Anti-railroad poster produced in 1839 by Philadelphia merchants.

Source: National Archives, Washington, D.C.

edge, the steamboat companies did all they could to prevent the erection of bridges across those barriers. Like any good anticompetitive strategy, however, the effort was cloaked in an assertion of the public interest. This time that interest was the safety of water navigation.

Two weeks after the completion of the first bridge across the Mississippi, on May 6, 1856, the steamboat *Effie Afton* crashed into the bridge. An engineering marvel that took three years to build, the structure between Rock Island, Illinois, and Davenport, Iowa, was rendered inoperable as one section was engulfed in flames from the burning boat.

The next boat behind the *Effie Afton* bore a banner that proclaimed "Mississippi Bridge Destroyed, Let All Rejoice." The banner, coupled with the nearby vessels' celebratory whistles and horns, suggested that, far from being an "accident," the crash was an escalation of the ongoing conflict between the watermen whose business was threatened and the railroaders (and through them a battle between the water city of St. Louis and the rail town of Chicago).[74]

The owners of the *Effie Afton* sued the owners of the Rock Island bridge, claiming a hazard to navigation. They sought $200,000 in damages, hoping the high cost of the damages would make bridges unprofitable and force railroads to continue to unload their freight on one bank, deposit it on the watermen's boats, and ferry it to the other side.[75] The bridge owners hired Abraham Lincoln as their lead attorney.

The trial before a local jury lasted two weeks. Lincoln's summation took two days. At the heart of Lincoln's reasoning was that a person had as much right to move across a river as he did to move up and down it.[76] "This [east-west] current of travel has its rights as well as that of north and south," Lincoln argued, "This bridge must be treated with respect in this court and is not to be kicked about with contempt."[77]

Lincoln's rationale may seem logically persuasive to us today, but the matter was such a revolutionary and emotional change, affecting such a major local institution, that the jury deadlocked. The nine jurors voting for the bridge and the three against it were a microcosm of the difficulties the new network was rapidly imposing on the stability and security of traditional lifestyles. Fortunately for Lincoln and the railroad, the deadlock allowed the bridge to remain. It was, the *Chicago Tribune* proclaimed, "virtually a triumph for the bridge."[78]

The railroads' assault on tradition was not lost on the congressional representatives from the southern states. The doctrine of states' rights, built on the protection of local institutions, kept southern rail lines short and sparse. Where there were railroads in the South they served their state, not the greater region. Most southern railroads stopped at the state line and did not connect with the lines of the adjoining state. As a result, at the start of the Civil War two thirds of the nation's rail mileage was north of the Mason-Dixon Line.[79]

When war broke out—a struggle between the interconnected industrial North and a southern system that clung to isolated independence, including the "peculiar institution" of slavery—the South's earlier unwillingness to embrace the new technology had military consequences that helped resolve the issue.[80] As the Union army used the North's vast rail network to its advantage, the South's inability to move men and equipment along interconnected routes proved such a disadvantage that Confederate general Robert E. Lee called for a program to connect the lines and thus improve the ability to shift troops from one area to another. His request fell on deaf states'-rights ears. The doctrine that had helped provoke the war helped to settle its outcome.[81]

In the North, however, the wartime absence from Congress of southern representatives had the benefit of eliminating the opposition that had prevented federal aid to expand the railroad westward. At the very depths of the fight to save the Union, Abraham Lincoln seized on this political opportunity to champion railroad expansion. On July 1, 1862, Lincoln signed the Pacific Railway Act, authorizing a government program to enable the Union Pacific Railroad to build west from the Missouri River and the Central Pacific Railroad to build east from Sacramento to create the first transcontinental railroad.

Lasting Legacies

The golden spike that united the ribbons of steel from the east and west in 1869 was a capstone on distance as the defining factor in mankind's destiny. Steam-powered speed meant that geography ruled no more. It had been a scant forty-four years between the first steam railroad linking Stockton and Darlington and the golden spike in the Utah wilderness. Never in human experience had such a transformational force been so rapidly imposed.

The transcontinental railroad was the final blow to the Jeffersonian ideal of a nation built on agricultural economics. A nation that was interconnected was also a nation that was industrialized, even on the farm, where production for newly broadened markets expanded landholdings and introduced mechanization. The ideal of agrarian independence yielded to industrial interdependence. That revolution, in turn, changed the nature of the limited government Jefferson and his colleagues had created, while at the same time demonstrating the genius of the flexibility built into their Constitution.

Government, which had enabled railroad growth with land grants and other policies, soon was forced to confront the need to balance the railroad's economic power. The first great grassroots political uprising grew out of the 1867 founding of the National Grange of the Patrons of Husbandry. In the following decades the Grange championed regulation of the railroads, particularly of their rates.

Because small agricultural communities rarely had more than one rail line, that company could extract what economists call "monopoly rents," rates that were as high as the traffic—and the market—would bear. A short haul from the farmers' small towns to an entrepôt such as Minneapolis, Chicago, Kansas City, or Omaha might cost twice as much per unit of weight as would a longer haul across a route with competitive carriers.

Through the Grange, farmers organized to offset the might of the railroads. Individually, a farmer was no match for the political heft of the railroads. Collectively, however, farmers could take on the big companies. Banding together, the users of the railroads harnessed their aggregate strength to counter corporate size. The Grange's early successes were in state legislatures, but those results were limited to specific states. In 1887, pressure from the Grange convinced Congress to pass the Interstate Commerce Act, creating the first federal independent regulatory agency, the Interstate Commerce Commission. This was the beginning of big government as a countervailing force to big business.[82]

At around the same time as farmers were banding together, workers were following a similar path to assert their rights as labor increasingly became a depersonalized input for corporate budgets. Working on the railroad was a "brutally dangerous occupation" that wore down workers as surely as it wore down rails, ties, and rolling stock.[83]

When in 1877 most eastern railroads cut wages (some more than once), workers responded collectively. The Great Railroad Strike of 1877 spread throughout the nation before being put down by federal troops. The die had been cast, however. The only way to get The Man's attention and respect was to act together, and Big Labor was born. To protect themselves against corporate behemoths, railroad men sought strength in numbers and common cause. They organized into four brotherhoods: engineers (1863), conductors (1868), firemen (1873), and trainmen (1883).

Following World War I, during which the rails were nationalized to preclude labor unrest, Congress passed a series of railway labor acts dealing with the rights of workers. Such legislation, which included the introduction of the eight-hour workday, became models for similar legislation affecting other industries' work environments.

The railroad companies, those whose goods they hauled, and the men who made up their workforce had ushered in a new era of big business, big labor, and big government.

As he launched the construction of the B&O Railroad, Charles Carroll compared his signature on the Declaration of Independence with the future he was inaugurating: "I consider what I have just now done to be among the most important acts of my life, second only to my signing the Declaration of Independence, if indeed it be even second to that."[84] Carroll's signature in 1776 had ushered in a new nation; his shovel fifty-two years later began a process that transformed that nation into the breadbasket of the world, the center of industrial expansion and innovation, and an economy shaping and shaped by a growing government.

We continue to live in the world the railroads made.[85] How we organize our cities to provide the necessities of public health, public safety, and public education all are an outgrowth of railroad-driven urbanization. The role of a centralized, corporate national government mimics that of the institutions it was expanded to oversee. The traditions and institutions that began with the railroad in the nineteenth century continue to hold sway over our existence almost 200 years later.

Connections

Just as the effects of railroad-driven changes continue today, the steam engines that drove that transformation can be found at the root of the computing engines driving even more rapid contemporary change.

"I wish to God these calculations had been executed by steam," a British mathematician declared in frustration at about the same time as the Stockton and Darlington Railway made its first run.[86] This was a vision that would dominate Charles Babbage's life and produce the precursor to today's computing engine.

Babbage was an acquaintance of the early railway pioneers and had been involved in many of the early debates about steam locomotion, including the appropriate spacing between tracks (the gauge). On hand for the inaugural run of the Liverpool and Manchester Railway in 1830, he proposed that the engines be fitted with what would become known as a cowcatcher to sweep obstructions off the line.[87] But it was Babbage's idea to harness steam power to a mechanical computing process that links him to today's networks.

In the summer of 1821, Babbage and a colleague were reviewing a set of mathematical tables being prepared for the Royal Astronomical Society. The calculations had been performed by "computers," individuals who did each calculation by hand. Because of the potential for error, it was common practice to have two different computers work on a project and to compare the results as a cross-check. Inconsistencies were flagged for recalculation.

Reviewing the tables was mind-numbing. One reviewer would read to the other a list of numbers as his partner compared them with the set he was studying. The intense concentration and repetitious boredom would cause the reviewers to stumble and have to go back and start again. It also caused Babbage to call for an automated solution.[88]

By the following June, Babbage had completed a small working model of a mechanical calculator that he called a "difference engine." A collection of vertical rods, each of which stored a single digit that was the result of the interaction of nine cogged wheels on the rod representing the numbers zero to nine, the difference engine was a monument to both mathematics and precision machining. Whereas the crude calculators of the time required human intervention (and thus the introduction of errors) to deal with the "carry" issue of moving a number from the units to the tens, from the tens to the hundreds, and so forth, Babbage's difference engine automated the "carry."[89] Charles Babbage had mechanized mathematics.

As he worked on continually improving the operation of the difference engine, Babbage was confronted by what he described as a "shadowy vision" of how "the whole of arithmetic now appeared within the grasp of mechanism."[90] He began to consider a much more complex apparatus, an "analytical engine" capable of feeding the result of a previous calculation back into the beginning of another calculation. He called it "the engine eating its own tail."[91] It would become the essential concept behind the modern computer.

Between the summers of 1834 and 1836, Babbage had conceptualized the components of what we today recognize as a computer. In incredibly detailed blueprints, Babbage laid out a central processing unit (which he called a "mill") that would rely on its own internal procedures, expandable memory (which he called the "store"), an input device driven by punch cards, and an output printer. The unit used base 10 arithmetic and was capable of if-then conditional branching, the preparation of a result in advance of its need (equivalent to microprogramming), and the use of multiple processors to speed the calculation by splitting the task (parallel processing).[92]

Charles Babbage built none of his engines. While he did turn out a small-scale difference engine, its full capabilities were never built because of disputes with his chief engineer and the cessation of funding from the British government. Nor was the analytical engine ever constructed. Components were built, but the entire apparatus was never tied together. Had it been, it would have been a monster the size of a small steam locomotive.

In 1991, however, the London Science Museum did what Babbage had not. Using the original blueprints and the technology of the mid-nineteenth century, the museum constructed an 8,000-part difference engine. It performed exactly as Babbage had forecasted more than 160 years earlier, producing error-free calculations.[93] The museum is currently constructing an analytical engine to Babbage's specifications.[94]

In the 1930s, multiple teams on both sides of the Atlantic labored to create the first digital computer. Amazingly, they were ignorant of Babbage's work and how their efforts had been preceded a century and a half earlier by the genius of one man and the excitement of the Age of Steam.

Four

The First Electronic Network and the End of Time

Charles Minot was impatient. Stuck on a siding at the Turner, New York, station, the railroad executive was anxious to continue his westward journey. The eastbound train, running on the single line of track, was behind schedule, however. The executive was stuck waiting . . . and waiting.

In the early days of railroading the challenge of managing trains headed in opposite directions over a single line of track was supposed to be solved by keeping to a schedule (called running "by the book"). The schedule had built into it a prioritization of trains (for instance, express over locals) and a safety buffer that gave each train a block of track time throughout its course. This meant that a train heading in the opposite direction could not encroach on that buffer and had to wait at a siding for the other train to pass. Should the other train not arrive within the appointed time, the rules allowed for the waiting train to proceed slowly behind a flagman on the lookout for the oncoming locomotive. When the two trains' conductors spotted each other, one would back up to the nearest siding and allow the other to pass.

When running by the book met Murphy's law, however, the system fell apart, and Mr. Minot was the wrong person to be inconvenienced by such a breakdown. The general superintendent of the New York & Erie Railroad, Minot had a reputation for being results-oriented, quick-tempered, and curt.[1] On this fall day in 1851 these quirks led him to do something that had never been done on an American railroad: he used the telegraph that ran alongside the tracks to communicate with stations down the line and direct the activity on the line ahead.

Minot telegraphed the next station, fourteen miles away in Goshen, New York, and asked if the eastbound train had passed. The answer quickly returned that the eastbound had yet to be seen. Rapidly, Minot sent another message, "To Agent and Operator at Goshen: Hold the train for further orders, signed, Charles Minot, Superintendent." He then gave the engineer on his train the order to proceed. Following protocol, it was a written order: "Run to Goshen regardless of opposing train."[2]

The engineer was far less enthusiastic than Minot about trusting the fate of his train to a few sparks on a telegraph line. He didn't care what the dots and dashes of the new technology said. He would not move his train until he saw the other train pass with his own two eyes.

Overruling the engineer, Minot took the controls himself. The anxious engineer retreated to the very last row on the very last car of the train, convinced it would soon cannonball head-on into the eastbound express.

When Minot successfully reached Goshen, the eastbound train still had not arrived. The same process was repeated. Minot telegraphed the next station to determine the eastbound's status. Minot received a quick response. Minot's train rolled forward. The same pattern followed at the stations after that. Not until the fourth station did the eastbound train finally appear.

Charles Minot had harnessed the only thing faster than a speeding locomotive to govern the movement of the iron horse.

Soon railroads and telegraph companies were sharing each other's assets. Commercial telegraph wires were strung along the track, and at each station railroad employees doubled as telegraphers. In return, railroad traffic took precedence on the wire and traveled for free. The local railroad station, which had been the town's physical link to the people and products of the outside world, now became the town's information center.

It was the breakthrough moment for the new telegraph network. "Of all the innovations which entrepreneurs, great and small, brought to the development of the telegraph industry," one historian observed, "none is more important nor dramatic than the discovery of the symbiotic relationship between the telegraph and the railroad."[3] As that symbiosis spread, the web of wires created what today we refer to as "telecommunications." It was the beginning of the infrastructure of the information age.

"Flash of Genius"

The same year as the first run of the Stockton and Darlington Railway (1825), a hero of the American Revolution, the Marquis de Lafayette, was on his fourth and final triumphal visit to the United States. As befitted a man who had been a colleague in arms of Washington and crucial in winning French support for the colonists, the city fathers of New York determined Lafayette's image should hang at City Hall alongside other giants such as Washington, Clinton, Jay, and Hamilton.

They commissioned the noted American artist Samuel F. B. Morse for the task. In early February, Morse departed to join Lafayette in the nation's capital. The trip from his home in New Haven to Washington was a four-day trek by road.

Settling into the orbit of the great man, Morse spent the evening of February 9, 1825, with Lafayette at the White House. In addition to visiting with President Monroe, the artist also chatted with President-elect John Quincy Adams, whose election had just been decided by the narrowest of margins in the House of Representatives. It was a heady evening. Eager to share his experience, Morse wrote his wife a lengthy account of the events. He closed the letter with a wistful "I long to hear from you."[4]

The addressee, Lucretia Morse, had died three days earlier.[5]

The following day Morse received a letter. "My Affectionately-Beloved Son," his father wrote, "My heart is in pain and deeply sorrowful, while I announce to you the sudden and unexpected death of your dear and deservedly-loved wife."[6]

The new widower raced home. He arrived several days after his wife's burial.

Still mourning his wife's death, four years later Morse departed for an extended, hopefully healing tour abroad during which he would study the painting techniques of the European masters. Returning home in October 1832 aboard the packet ship *Sully* he was struck with what he immodestly called a "flash of genius."

One night the *Sully*'s dinner conversation turned to the relatively new concept of electromagnetism. "I then remarked," Morse later recorded, "if the presence of electricity can be made visible in any desired part of the circuit I see no reason why intelligence might not be transmitted instantaneously by electricity."[7]

It was an unexpected insight from a man whose activities had hitherto been focused on art rather than science. Nevertheless, the six-week sea journey gave Morse plenty of time to fill his artist's sketchbook with notes and diagrams about the issues that needed to be solved.

Despite his not-so-humble "flash of genius" assertion—and an effort in his later years to gather endorsements from his fellow passengers establishing his parentage of telegraphy—Morse understood little about the physics surrounding the transmission of electric energy. He was also unaware that the idea for an electric telegraph was almost eighty years old and had already seen multiple demonstrations.

In many ways Morse's ignorance acted to his advantage. Assuming away the problems of practical physics, he focused instead on the input and output mechanisms. From the outset, Morse believed there needed to be a permanent record of the message. His other conclusion aboard the *Sully* was that the message should be expressed in digits rather than letters as there were only ten digits but twenty-six letters (a number increased by accompanying capitals, punctuation, and other marks). Morse conceived of combining bursts of electric energy, expressed as dots and dashes, to represent the numbers he would transmit. He even began to play with the best permutations of such dots and dashes.[8]

To turn the transmitted numbers into words, Morse envisioned a codebook that assigned a numerical representation to every word. To compose a message, one simply looked up the words and transmitted their numbers. At the receiving end the process was just the reverse.

To compose these signals, Morse inserted metal teeth into a slot cut into a yard-long piece of wood. He called it a "port-rule." A crank then drew the sawtooth-embedded stick under a lever that would ride up and down on the sawteeth. At the other end of this armature was a connection to a battery so that when an "up" motion along the teeth was produced, reciprocal movement at the other end of the lever completed the circuit. Likewise, every "down" motion broke the circuit. The impulses corresponded to the numbers one through ten. The number 1, for instance, was a single tooth; 2 was a tooth, then a short space, then another tooth. The number 6 began to introduce long spaces and was a single tooth followed by a long space.[9]

At the other end, Morse designed a moving roll of paper and a pen or pencil attached to an electromagnet that hung over the paper like a pendulum. He called this a "register." As the current (or lack thereof) forced the writing instrument to move from one side of the paper to the other, it left marks in the shape of a V that corresponded to the impulses generated at the originating station. The number 23, for instance, would appear as VV VVV.[10]

Not a New Idea

Disembarking from the *Sully* in New York on November 16, 1832, Morse told the ship's commander, "Well Captain, should you hear of the telegraph one of these days, as the wonder of the world, remember the discovery was made on board the good ship *Sully*."[11]

Morse's grandiose proclamation notwithstanding, like all great network breakthroughs, the telegraph was the result of the gradual convergence of other technologies. It began with the almost-mystical transmission of electricity through a wire.[12] In 1729, in London, Stephen Gray hung wire from silk threads and transmitted a friction-generated current from one end to the other.[13] As early as 1753 a letter from someone identified only as "C.M." proposed sending messages over such a wire by means of electrical pulses. Published in *Scot's Magazine*, the idea was described in an article titled "An Expeditious Method of Conveying Intelligence."[14]

The concept put forth in the *Scot's* piece—a separate line for each letter of the alphabet that would trigger a signal at the receiving end—was demonstrated in Geneva twenty-one years later by the French inventor Georges Le Sage.[15] A bizarre application of the same concept was put forward in 1804 by the Catalan scientist Don Francisco Salva i Campillo, who proposed that the end of each wire be attached to a person, who would shout out the assigned letter when he or she received a shock.[16]

Morse was also ignorant of the electromagnetic signaling systems that had been constructed on the American side of the Atlantic. As early as 1827, a rudimentary demonstration project had been built on Long Island.[17] Credit for the first electric telegraph, however, goes to Professor Joseph Henry. In January 1831, the year before Morse's "flash of genius," Henry described the concept in an article in Benjamin Silliman's *American Journal of Science and Arts* (whose title was later shortened to *American Journal of Science*). He then went on to build a demonstration that used electricity to command an armature to strike a distant bell; it was precisely the functionality of a telegraph with a circuit being opened or closed at one end to produce a desired action at the other.[18]

The idea of moving information at speed by disembodying it from the physical was well established. Signaling over distance by smoke or fire had been around for centuries. The English term "telegraph" was derived from the French *télégraphe* ("far writing), the name Frenchman Claude Chappe gave to a system of optical relays he had developed.

Early fire or smoke signals were limited in that they could only convey broad contextual information (for instance, "enemy sighted" rather than "100 cavalry passed at 4:00 heading southwest"). On March 2, 1791, Chappe overcame that limitation by successfully transmitting a nine-word message over ten miles in four minutes.[19]

Chappe's first telegraph used panels painted black on one side and white on the other. The signal was sent by the combinations of the panels. Later on, he would adopt a semaphore-like system. The French Revolution inconveniently intervened and revolutionary paranoia made suspect any new means of sending messages. But when Napoleon came to power in 1799 he embraced the concept and ordered the construction of a string of Chappe's towers. Like Morse, Chappe's signals were recorded in a codebook.[20]

The Chappe idea was soon replicated throughout Europe. In the United States similar systems sprang up, principally around seaports as a means of reporting approaching ships. The Telegraph Hills in Boston and San Francisco are the legacies of such systems—high points from which distant activity could be observed by telescope and which themselves could be observed as the originating point of optical signals conveying the information.[21]

By 1837, the British scientist Charles Wheatstone and his partner, William Cooke, had improved on the early ideas for electronic communications and were sending messages over the mile and a half between London's Euston and Camden Town railway stations. Their cumbersome system required five wires, each connected to a row of five needle-like arrows in the middle of a diamond-shaped grid. By providing current to the appropriate line the originator could send electricity through a coil wrapped around one of the five arrows at the other end, making it pivot to the right or left.[22] The appropriate letter of the alphabet was indicated by two arrows pointing toward it.

Not only was the Wheatstone-Cooke device cumbersome, but it could only signal twenty of the alphabet's twenty-six letters (missing were *c*, *j*, *q*, *u*, *x*, and *z*). Nevertheless, it was transmitting messages via electronic impulses. The British government gave Wheatstone and Cooke a patent for their invention.[23] The Wheatstone-Cooke system went into commercial service in 1838 alongside the tracks of the new Great Western Railway out of Paddington Station.

In February 1837 (the same year as the Wheatstone-Cooke demonstration in London), U.S. treasury secretary Levi Woodbury, responding to a congressional directive, issued a request for information on the feasibility of a "system of telegraphs." The inquiry was about a Chappe-like system, and seventeen of the eighteen respondents described just such an optical signaling system. Samuel Morse's response reimagined the undertaking as an electronic signaling system.[24]

In his rooms at New York University, Morse, now a professor of the literature of the art of design, had been painting, teaching, and experimenting. He developed his ideas from the *Sully* using whatever materials he could get his hands on. An early register, for instance, was an artist's canvas stretcher, nailed to a table, from which was suspended the electromagnet.[25] In his response to Secretary Woodbury, Morse reported on his tests and promised to keep the other man informed.[26]

One of those subsequent reports was of the successful transmission of signals through ten miles of spooled wire. Bullish on this development, Morse confidently predicted that "we have now no doubt of . . . effectuating a *similar result at any distance*."[27] Here was the true breakthrough in telegraphy: overcoming the resistance of the wire, which, as distance increased, attenuated the strength of the signal until it eventually disappeared. The technology that overcame this problem did not belong to Samuel Morse, however. Morse went to great lengths to deny such precursors ever existed and to claim that he and his "flash of genius" were the sole parents of telegraphy. But that was not the case.

Six years before Morse's letter to the treasury secretary, Joseph Henry's article in Silliman's journal had identified the *intensity* of electricity (that is, the voltage) as more important than the *quantity* (that is, the current). Shortly prior to his letter to Woodbury, Morse had only been able to transmit over relatively short distances, not the distance that practical telegraphy would mandate. Another NYU professor, Leonard Gale, told Morse about Henry's article and how voltage could be increased by using a multi-cell rather than a single-cell battery for power. Gale also shared Henry's observation that the power of an electromagnet could be increased by increasing the number and tightness of its wire wrappings. With these two modifications, the reach of Morse's telegraph jumped from forty feet to ten miles.[28]

It was these breakthroughs that allowed Morse to brag to Woodbury about "a similar result at any distance." The following year (1839), when Morse was struggling with even greater distances, he would go see Professor Henry and walk away with another Henry revelation to mitigate power loss over great distances.

At the point on the line where the wire's resistance had reduced the flow of electricity to almost nothing, Henry installed an electromagnet. Receiving just enough of the dying current to activate it, the electromagnet closed a new circuit connected to a new power source. The resulting new and powerful signal, a perfect clone of the original, then continued down the wire until it, too, weakened and was reamplified by the next relay.

Unlike Samuel Morse, Joseph Henry was an unassuming man of science. It was Henry's belief that scientific advancements should be freely available for the betterment of mankind rather than the enrichment of one man. As a result, he did not patent any of his discoveries. Morse did not suffer from such selflessness; he incorporated Henry's ideas without as much as a nod of recognition. When it came time for Morse to patent "his" ideas in 1840, the patent included the technology that Henry had selflessly declined to claim for himself.

In later years Joseph Henry—who went on to become the first secretary of the Smithsonian Institution and the unofficial science adviser to President Lincoln—would dismiss Morse's grandiose efforts to "claim for myself

. . . the invention of a mode of communicating intelligence by electricity."[29] "I am not aware that Mr. Morse ever made a single original discovery in electricity, magnetism, or electromagnetism, applicable to the invention of the Telegraph," the scientist would testify.[30]

Morse Goes to Washington

While Morse may not have been a paragon of virtue, he was without a doubt the hardest-pushing promoter of the still seemingly miraculous concept of an electronic means of leaping both time and space. What Morse lacked in scientific knowledge he more than made up for in bravura.

Morse took his telegraph on the road. In early January 1838 the Morristown, New Jersey, *Jerseyman* reported on a demonstration in that city; "Professor Morse's Electromagnetic Telegraph," the headline announced. The following month Morse and his colleague-cum-assistant, Alfred Vail, were in the nation's capital to demonstrate their device.

The Committee on Commerce of the House of Representatives was the site of the telegraph's grand unveiling. Morse and Vail rolled two five-mile spools of wire into the meeting room, and members of Congress came to view the demonstrations. President Martin Van Buren and cabinet members journeyed from the other end of Pennsylvania Avenue. Morse asked Van Buren to whisper a message to be transmitted. Out the other end came "The enemy near."[31]

At a time when electricity was a vague scientific concept (Edison's light bulb wouldn't appear until 1879), Morse's demonstration of his telegraph was awe-inspiring.

"The world is coming to an end," Vail reported one attendee asserting.[32]

"Time and space are now annihilated," another visitor correctly observed.[33]

Such heartfelt sentiments from such powerful individuals, however, do not a revolution make. On April 6, 1838, the Commerce Committee reported a bill appropriating $30,000 for a fifty-mile trial of the technology it had witnessed. The committee's report described the telegraph as "a revolution unsurpassed in moral grandeur by any discovery that has been made in the arts and sciences."[34]

The bill went nowhere. The momentum Morse seemed to have generated with his February demonstrations dissipated as elected representatives came to grips with the political consequences back home of a large appropriation for a project beyond the comprehension of most of their constituents. It was a time of domestic austerity following the Panic of 1837. Casting one's vote for what seemed a parlor trick was deemed to be a politically unhealthy act.

Once again, Morse's drive for recognition caused him to behave questionably. Unbeknownst to the other members of Congress, the principal sponsor of the funding legislation was on the take from Morse. Between the February demonstration and the April committee report Morse had secretly given a one-quarter ownership position in his technology to Representative Francis O. J. Smith of Maine, the chairman of the Commerce Committee. Yet, even with such a powerful fixer, voting for the seemingly harebrained idea of messages by sparks was too politically risky for a majority of his colleagues.

It would be five years before Congress would again revisit the telegraph issue. During that period Morse traveled to Europe and was unsuccessful in acquiring foreign patents and attracting investors. In the process, he learned his "flash of genius" had occurred to others. The Wheatstone-Cooke team was ahead of Morse in the successful implementation of their invention. After successfully installing their telegraph for use by the Great Western Railway in England, the two were exploring bringing it to the other side of the Atlantic.

The unsuccessful European trip disheartened Morse and allowed other interests to take center-stage in his life. While pitching his ideas in Paris, he had met Louis Daguerre and had seen his early photographic technique. Returning home, Morse combined his artistic proclivities with the new process and set up a studio for what he called "photographic paintings."[35] He also ran and lost the 1841 race for mayor of New York on an anti-immigrant, anti-Catholic platform.

Amazingly, Samuel Morse had yet to patent his telegraph ideas. He had received a "caveat" from the Patent Office in 1837—essentially a placeholder, establishing the claim's priority in front of others, but lacking any specifics about the technology's processes. It wasn't until June 20, 1840, that patent no. 1647, "a new and useful Improvement in the mode of communicating information by signals, by the application of Electro Magnetism" was granted.[36]

Congressional infatuation with message signaling stirred again in 1841. The recent congressional elections had given the Whigs and their platform of governmental promotion of internal improvements a majority in both houses. Once again, optical signaling was under consideration. Sitting atop the Capitol building (which was then without its current dome) was a semaphore, part of a trial optical system Congress was considering funding. Morse hired a lobbyist, Isaac Cohen, to advocate on behalf of an electric telegraph. Cohen failed.[37]

When the third session of the Twenty-Seventh Congress returned in December 1842 for a final three-month meeting, Samuel Morse was present as his own lobbyist. His illicit partner, Representative Smith, was gone, however; he had not stood for reelection. In his place Morse found a new (and legitimate) ally in Representative Charles Ferris of New York City, a member of the Commerce Committee.

A telegraph line was built from the House Commerce Committee room to the Senate Naval Affairs Committee room. Once again, Morse gave demonstrations to any and all. Not everyone was awestruck. "I watched his countenance closely, to see if he was not deranged," wrote Indiana senator O. H. Smith, "and I was assured by other Senators after we left the room that they had no confidence in it."[38]

Morse was desperate and broke. At one point toward the end of his time in Washington, Morse calculated that after buying a train ticket home to New York, he would have thirty-seven and a half cents to his name.[39]

Finally, on December 30, 1842, the Commerce Committee reported Representative Ferris's bill (H.R. 641) to the full House. Attesting to the interest in the topic, the House ordered 5,000 copies of the committee's report printed.[40] The representatives were undoubtedly anticipating the need for something to explain the seemingly wild idea to their constituents.

First, however, the idea of messages by lightning had to survive the doubters in Congress. On February 21, 1843, the full House of Representatives considered the recommendation of its Commerce Committee that $30,000 be appropriated to "test the Practicality of establishing a System of Electro-Magnetic Telegraphs by the United States."[41] The debate turned into a carnival.

Representative Cave Johnson of Tennessee was the ringleader of the opposition. He "raved and scolded, and ranted, and screamed, and foamed against the House, like a demonically possessed man," one newspaper reported.[42] Johnson even moved to amend the bill to appropriate half the funds to study sending messages by mesmerism (hypnotism). Members' laughter and witty ripostes carried throughout the chamber. As Morse sat in the gallery, a witness to the mockery, one congressman objected to the folderol and requested that the mesmerism amendment be ruled out of order. Joining in the levity, the presiding officer ruled, to more laughter, that absent "a scientific analysis to determine how far the magnetism of Mesmerism was analogous to that employed in telegraphs," the amendment was in order. After the representatives had their fun, the amendment was defeated, but not before garnering twenty-two votes.[43]

Ultimately, two days later on February 23, the House of Representatives passed the appropriation by the razor-thin margin of 89 ayes to 83 nays. Seventy congressmen chose to abstain rather than have to make a decision one way or another.[44]

Passage by the House was a victory, but time was running out. There were only eight days remaining before the end of the Twenty-Seventh Congress. Morse turned his attention to the Senate. On March 3, the last day of the session, Samuel Morse despondently watched from the gallery as the Senate moved through its agenda. His allies on the floor had told him not to be too hopeful. With the carnival atmosphere of the House's debate just a few days past, one could only imagine what could happen on the Senate floor; any delay at this late hour would be fatal.

A miracle of miracles occurred, however; as the Senate droned on late into the night, the appropriation was finally raised. The official record of the action is terse and without the mischievousness of the House: "The House bill making appropriations to test the plan for electro-magnetic telegrams, was read the third time, and passed."[45] There was no dissent. That evening President Tyler signed the bill into law.

Mr. Morse would have his telegraph.

But Will It Work?

The $30,000 appropriated by Congress equates to approximately $1 million today.[46] Anointing himself with the grandiose title of superintendent of the electro-magnetic telegraph, Morse drew funds from the secretary of the treasury and prepared to begin construction. Achieving his position with no scientific expertise, Morse clearly expected he could oversee a complex construction project with no management expertise either. The absence of both skills would become manifest.

Construction began three weeks later than planned, on October 21, 1843, at the Baltimore end of the line. The fifteen-year-old Baltimore & Ohio (B&O) Railroad consented to allow the wires to follow its right-of-way to Washington.[47] Fearful of vandalism, and bowing to his chauvinistic beliefs in evildoers from abroad, Morse decided to bury the wires inside flexible lead tubes that were specially made for the purpose. To oversee construction Morse appointed his patent partner, former representative F. O. J. "Fog" Smith, to procure the conduit and wire as well as to contract for its trenching. Smith promptly gave the trenching contract to his brother-in-law, who in turn subcontracted it to a plow salesman named Ezra Cornell.

Once outside the grounds of the Baltimore rail depot Cornell's specially designed trenching apparatus went to work. As a team of eight mules pulled, the plowshare bit into the earth to create a narrow slit. A spool of lead tubing containing insulated wire mounted atop the plow fed into the trench behind the blade. As the plow progressed the walls of the trench collapsed to cover the wire.[48]

After advancing about ten miles Morse's construction began to collapse like the walls of Cornell's trench. The buried tubing was leaking. Morse turned to a new tubing supplier, but its product was defective as well. It was December and winter was setting in.

Morse had planned to have his project completed by the time Congress reconvened in December. Instead, December found him stuck about ten miles outside of Baltimore at Relay, Maryland. Worst of all, the part of the trial that had been built failed because the leaks in the conduit grounded out the signal.

It wasn't just technical problems that plagued Superintendent Morse. His erstwhile partner Fog Smith continued in the less-than-reputable behavior he had demonstrated in Congress. The supplier for the second batch of lead tubing charged $1,172 less than what was budgeted compared with the pricing of the earlier supplier. True to his ethical fluidity, Smith proposed that they not share this information with their government funders and that he and Morse quietly split the savings between them.[49] When Morse refused, relations between the two soured.[50]

Morse may not have been willing to engage in larceny, but a little deception was acceptable. He was regularly reporting to the treasury secretary as he drew new funds. At the same time, the newspapers covered his progress. Word could not leak out that thus far the project was a failure. Morse needed a plausible excuse to stop work and reassess the situation.

Ezra Cornell, who knew of the technical difficulties, provided the cover story. He proposed to Morse that his trenching tool have an accident. With Cornell driving, the device unexpectedly steered into a large rock, and the blade broke into pieces.[51] Morse announced the project would go into winter quarters, which would give him time to repair the trenching tool and test the line that had been laid thus far.

Those winter quarters were in Washington. Morse took up residence in the home of an old school chum, Commissioner of Patents Henry

Ellsworth. The commissioner also gave him a room in the basement of the Patent Office to store his materials. During the winter hiatus Morse's faithful assistant Alfred Vail and Ezra Cornell hit the books. Unlike their boss, they determined to accept that Morse had not had an isolated flash of genius, and to strive to learn from the experiences of others. Vail discovered an English publication reporting how Wheatstone and Cooke had also had trouble with buried wires and had ultimately opted to suspend the lines above ground. Vail, once again playing a determinative role in the success Morse would claim for himself, convinced Morse that aerial construction was the only alternative.

The project was nearly broke. By December, Superintendent Morse had spent $23,000 of the $30,000 the Congress had appropriated. All he had to show for it was a ten-mile stretch of wire that didn't work. Vail wrote to his wife, "I should not at all wonder if the appropriation is exhausted before we are able to do a thing."[52] Down in the basement of the Patent Office, Morse's team spent the winter tediously reclaiming the wire from the spools of lead tubes, unwrapping its insulation and repurposing it to be hung from poles. To replenish his dwindling accounts Morse sold the lead tubing as scrap.

Vail may have discovered that Wheatstone and Cooke suspended their wire, but there remained the problem of how to keep the line from grounding out when it came in contact with the pole. The inspiration came from the glass knobs on the dresser in Ezra Cornell's hotel room. Because glass does not conduct electricity, anchoring the wire to glass insulated the current. The ultimate solution, glass insulators, bore an uncanny resemblance to their furniture counterpart.[53]

If the technical problems weren't sufficient to provoke ire, the decision to cease trenching enraged Fog Smith. Unbeknownst to Morse, Smith had bought a half interest in Cornell's trencher. Not only was his brother-in-law no longer needed for the trenching, Smith's dream of riches from licensing Cornell's device to future telegraph companies was destroyed. Smith's relations with Morse, already tense, broke into open warfare. Morse began to communicate with his partner only in writing.[54]

Nevertheless, in mid-March 1844, the Morse team began boring holes approximately every 200 feet along the B&O line. Into these holes went thirty-foot poles with the bark still on them, with a crossarm atop the poles that carried the wires. When the wire came to the crossarm it was wrapped in gum-shellac-coated cotton and fitted between glass insulators.

After the line had progressed seven miles (this time with Washington as the starting point), Morse ran some tests. It worked! Every day the line was extended and then tested. Every day set a new American record for the longest transmission over an operational telegraph line.

By May 1 the line was halfway to Baltimore, where the Whig party was meeting to select its presidential candidate. Washington waited with bated breath for the first train from Baltimore carrying the news of the convention's decision. Ever the showman, Morse saw a promotional opportunity. He had Vail meet the Washington-bound train at Annapolis Junction, the furthest point of the telegraph line. Learning the results of the convention from those aboard the train, Vail immediately telegraphed the news to Morse in the capital. By the time the train reached Washington an hour and a quarter later, the news had already spread throughout the town that Henry Clay and Theodore Frelinghuysen were the Whig nominees.

Amid all the trials and travails of his project, Samuel Morse fell in love. The fifty-three-year-old Morse was smitten with Anne Ellsworth, the eighteen-year-old daughter of Patent Commissioner Ellsworth, with whom he was lodging. "I desire sincere love to dear Annie," he had written her father in February. Alfred Vail feared Morse's preoccupation with romance was interfering with his project. "The secret is," Vail wrote his brother, "[Morse is] so much in love he doesn't know what he is about half the time."[55]

The romance appears to have been one-sided, but it produced one of the most memorable lines in history. Morse asked Anne to suggest the text for the first official transmission over the completed telegraph line. She chose a selection from the Bible, Numbers 23:23, "What hath God wrought!"

On May 24, 1844, Samuel Finley Breese Morse sat in the Supreme Court chamber on the east side of the Capitol building and tapped out Anne

Ellsworth's phrase to Alfred Vail forty miles away in Baltimore. Vail then sent a confirming message back to the Capitol.

At both ends the message was recorded as dots and dashes on a piece of moving paper tape. Once again, Alfred Vail had introduced an integral innovation for which Morse would be given credit. In place of the crude portrule and its saw-toothed parts that Morse had designed, Vail introduced the telegraph key. At the other end, where Morse had envisioned a cumbersome register using a suspended electromagnet pendulum, Vail devised a machine that embossed the dots and dashes on a paper tape.

Over the years there has been much confusion about the punctuation of the first telegraph message. The message was commonly believed to be an interrogatory statement that closed with a question mark.[56] That, however, is not how the phrase appears in scripture. In Numbers the phrase closes a passage of exultation about Jacob and Israel with an exclamation point. Morse transmitted the message without punctuation as "What hath God wrought." After the message was transmitted it was transcribed, at which time a question mark was added, even though one was not transmitted.[57]

The historic message also illustrated how Morse and Vail had yet to realize that the code could simply be translated by listening to the dots and dashes on the telegraph sounder. In the device used for the first message, the sounder made marks on a tape that was fed through it. The letter *W*, etched on the tape as a dot followed by two dashes, was followed by four dots for *h*, a dot and a dash for *a*, and a dash for *t*.

Three days after the immortal message, Morse's showmanship was again on display. Again, it was news from a political convention held in Baltimore. The Democratic convention was a hotly contested event. When delegates deadlocked between Martin Van Buren and Lewis Cass, the convention turned to James K. Polk on the ninth ballot. With the line now extended to Baltimore, Vail provided a play-by-play report to those gathered in the Capitol.

The following day, May 28, the telegraph determined Polk's running mate. In an effort to reach out to Van Buren supporters, the convention nominated New York senator Silas Wright. As was typical for the period,

Wright was not present at the convention, preferring to stay in the capital. When Vail telegraphed Morse the news, Morse told Wright. The senator, however, did not want the nomination. He asked Morse to telegraph his refusal to accept the convention's decision. When Vail delivered the news, the delegates were stunned. Not only had it only been a short time since the nomination, but the honor was being rebuffed from afar! The convention sent a telegram back asking Wright to reconsider. Wright replied that his decision was final.[58]

Perhaps the greatest miracle, however, came from Representative Cave Johnson of Tennessee. It was Johnson who had authored the mocking mesmerism amendment to the telegraph appropriation bill in the House. After seeing the back-and-forth negotiations with Baltimore, the onetime nemesis approached Morse and volunteered, "Sir, I give in. It is an astonishing invention."[59]

Spreading Like a Virus

Within a year, Cave Johnson was back in the middle of the telegraph debate. In March 1845 Congress transferred control of the Washington-Baltimore line from the Treasury Department to the Post Office Department. That same month President Polk installed Cave Johnson as postmaster general. Morse's former bête noire was now in charge of his project.

On April 1, 1845, Postmaster Johnson instituted a tariff for sending messages on the line he now controlled. Every four characters transmitted would cost one cent. The fees, however, were insufficient to cover the line's operating costs. At the end of its first six months the tariff had generated $413.44 in revenue while running the line had cost $3,284.17 in expenses.[60]

The technological success of the telegraph failed to drive revenue simply because Americans could not imagine how they could benefit from the breakthrough. Relying on his showmanship, Morse demonstrated the telegraph's capability to enable instantaneous communication over distance by staging chess matches between players at opposite ends of the line.

Cave Johnson, the recent convert, did not suffer from such myopia. While he was not convinced the telegraph would ever be a profitable enterprise, the postmaster nevertheless wanted the government to control it. "The use of an instrument so powerful for good or evil," he wrote in his first annual report to Congress, "cannot with safety to the people be left in the hands of private individuals uncontrolled by law."[61] It was an amazing conversion by the man who had once mocked the idea of instantaneous messages on the floor of the House of Representatives.

Understanding Johnson's vision, however, was a stretch for other policymakers. President Polk was generally opposed to such governmental expenditures on internal improvements. Even more telling were the remarks of Senator George McDuffie of South Carolina. The national leadership simply could not conceptualize what Morse and Johnson were promoting. "What was the telegraph to do?" McDuffie asked. "Would it transmit letters and newspapers? . . . And besides, the telegraph might be made very mischievous, and secret information . . . communicated to the prejudice of merchants."[62]

Where the government feared to tread, however, investors and entrepreneurs saw opportunity. One of Cave Johnson's rationales for government ownership was that private telegraph lines were already being built and the opportunity for the government would soon be lost. As Morse sold territorial licenses to his patent, the new licensees formed companies, raised capital, and built and operated telegraph networks. By 1851, more than fifty separate telegraph companies were operating in the United States.[63]

In a pattern similar to printing (and followed by the internet a century and a half later), early adoption of the telegraph was slow, only to subsequently grow rapidly and abruptly. The forty miles of the 1844 Washington-Baltimore test line had gradually expanded to 2,000 miles of telegraph wire by 1848. The next two years saw that mileage increase sixfold to more than 12,000 miles. Two more years (1852) and the mileage had almost doubled again. By 1860, the estimated telegraph mileage in the United States exceeded 50,000 miles.[64]

Amid this rollout of commercial telegraph service, the worm of doubt infested the public's understanding and acceptance of the new network. As he had with the railroad, Henry David Thoreau bemoaned the change. "We are in great haste to construct a magnetic telegraph from Maine to Texas," he wrote in *Walden,* "but Maine and Texas, it may be, have nothing to communicate."[65]

Thoreau and the masses were being thrust into a world they neither sought nor understood. The mid-nineteenth century was the most intense period of network-driven change the world had ever experienced. First, the growth of railroads demolished distance, prompting one writer to observe, "Space is killed by the railways and we are left with time alone."[66] Then the telegraph abolished time by somehow harnessing lightning. The comfortable patterns of centuries eroded seemingly overnight. "Try to imagine," one commentator observed, "the ambivalent anxieties of a freewheeling people with one foot in manure and the other in the telegraph office."[67]

As with earlier (and later) technological innovations, those anxieties were expressed in many surprising ways, including concern about the telegraph's impact on personal safety. If lightning was dangerous, then wasn't captured lightning more so? When a trial line was proposed in New York City in 1844, it was opposed on the grounds that moving lightning through wires across rooftops could itself attract lightning.[68]

What could not be understood had to be explained in terms of its otherworldliness. In a situation reminiscent of Johannes Fust's trip to Paris with the new printed Bible, clergy at the Baltimore end of the trial line determined its instantaneous messages could only be black magic. Morse's operator used the telegraph to tell his colleagues in Washington of the growth of this sentiment and warn about continuing the test. "If we continue we will be injured more than helped," he telegraphed.[69]

Confirming the predisposition to otherworldliness, one newspaper described the telegraph as "an almost supernatural agency."[70] As people tried to comprehend the phenomenon of disembodied electronic messages, other "telegraphic" interpretations took hold. Mesmerism, electrophysiology, and reformist Christianity blended to create a new popular spiritualism. Since the telegraph had leapt the boundaries of space, the spiritualists argued, it must also be possible to traverse the void of death. The belief caught on that a "spiritual telegraph" could be accessed through séances to connect the living with the dead.[71]

The End of Time

American society may have had "one foot in manure and the other in the telegraph office," but it was inexorably moving from the former to the latter.

Until the telegraph, "it took so long to obtain information that people lived their lives and made their decisions more in its absence than its presence."[72] The separation of information from its physical delivery was the first step in the information age. While the railroad had accelerated the delivery of information, the telegraph had separated information from its physical package.

The second message over the Washington-Baltimore line heralded what was to come. "Have you any news?" Alfred Vail tapped out to Samuel Morse.[73] The nature and uses of perishable information—events, financial news, or the coordination of business activities—would be forever altered by the telegraph. "The telegraph may not affect magazine literature," *New York Herald* editor James Gordon Bennett observed, "but the mere newspapers must submit to destiny and go out of existence."[74]

Instead of submitting to such a fate, however, newspaper editors like Bennett recast the nature of their offering. In the decade prior to the telegraph, Bennett and his ilk had transformed the newspaper from a partisan political rag directed at the moneyed class into the so-called "penny press," affordable to the working person. The penny press carried local news (much of it sensational) and was supported by advertising. The telegraph expanded the scope of these newspapers, enabling the delivery of information long before it arrived in the mail. In a world where timely information from afar was the most precious of luxuries, the telegraph made timely information both commonplace and essential.

Only four years after Morse's trial line, in 1848 six New York City newspapers, including Bennett's, set aside their fierce competition in order not to "submit to destiny and go out of business" but rather harness the new technology to shape a new destiny. Together they would share in the cost of distant correspondents and the delivery of their stories by telegraph. They called their enterprise the New York Associated Press. The most compelling content of each of the individual newspapers became the columns headed "News by Magnetic Telegraph" or a variation thereof. Within a few years the cooperative was providing telegraph-delivered news feeds for a fee to newspapers across the nation and had shortened its moniker to the Associated Press.

News by telegraph even changed journalistic style. Prior to the wire, newspaper stories were written in a narrative literary style. The telegraph instituted a more modern style in which important content is presented in the lead paragraphs, with less important color information following. This new style, which remains in use today, allowed the recipient newspaper to trim the story to fit the available space without losing the headline content.

The news that was of the greatest value, of course, was that with the potential to move markets. Among the first business institutions specifically created by the telegraph were the seven commodity exchanges that sprung up between 1845 and 1854. In Buffalo, Chicago, Toledo, New York, St. Louis, Philadelphia, and Milwaukee, interconnected traders created markets for wheat, corn, oats, and cotton.[75]

It wasn't just financial markets the telegraph enabled. As the railroad reduced the cost of transporting products, the telegraph allowed the coordination of the allocation of those products. An economy previously characterized by small firms operating in a local market was suddenly knit together. The ability to coordinate supply and demand at the regional and even national level enabled national-scale production and distribution.

Nowhere did this industrial coordination have a bigger impact than on the railroads along which the telegraph wires hung. As we saw with Charles Minot, most American railroads were single-tracked to preserve capital and cut construction time. As rail traffic grew, however, single-track lines created bottlenecks. The telegraph managed around those bottlenecks by coordinating activity along the entire line. By one estimate, the ability to maintain single-track lines in 1890 using the telegraph saved in track alone two and a half years of steel production. It has been calculated that the cash this saved was only half the total operational savings attributable to telegraphic coordination of rail activity. One historian described such savings as "perhaps the single largest unambiguous instance of the economic payoff of the telegraph."[76]

As the telegraph connected supply and demand to build regional and national markets, it also paved the way for the creation of antimarket monopoly forces. The telegraph became an economic Janus. At the same time that it was creating new markets through rapid communications, it was shrinking the competitive marketplace by allowing large firms to integrate and assume market-controlling activities for themselves.[77] As the railroad hauled raw material to a central site for mass production into products that it then redistributed to an interconnected market, the telegraph was the management tool that coordinated activities throughout the process. By thus creating economies of scale, the integrated and interconnected corporations gained a market advantage over smaller firms that ultimately put those firms out of business.[78]

Amid this economic upheaval, the technology that was knitting the nation together was also contributing to its dissolution. As new networks vaulted over geographic distances, they voided the physical separation that had shielded local idiosyncrasies. No group was more sensitive to this reality than the political representatives from southern states practicing the "peculiar institution" of human slavery. With the Mexican War under way in 1846, Congress considered legislation to fund a telegraph line from Washington to New Orleans to more rapidly communicate with those in the conflict. The proposal's rejection illustrated the continuing southern animosity to such means of interconnection. States' rights champion Senator John C. Calhoun, whose home state of South Carolina wouldn't even be touched by the line, led those who blocked it by challenging the constitutionality of a federal right-of-way crossing the boundaries of sovereign southern states.

Half a dozen years later, the Census Report of 1852 featured a dozen pages heralding the expansion of the telegraph, including a map of the existing telegraph lines. North of the Mason-Dixon Line the network looked like a spider's web. South of that demarcation, however, were only two threads, one running down the East Coast and the other down the Mississippi Valley.

Mr. Lincoln's T-Mails

When in 1860, as a part of his campaign for the Republican presidential nomination, Abraham Lincoln journeyed to New York City, he was not only reaching out to eastern Republicans but also exploiting the advantage the telegraph had bestowed on New York as the nation's news hub.

Lincoln's speech at Cooper Union, the institution endowed by Peter Cooper, builder of the *Tom Thumb* locomotive and an early investor in the expansion of the telegraph, sped over the wires to become national news. In a conscious effort to create what today is called a sound bite, Lincoln concluded his speech with "Let us have faith that right makes might, and in that faith, let us, to the end, dare to do our duty as we understand it." His strategy worked: the crisp, direct, and powerful formulation encapsulated his message and was sped by telegraph across the nation.

Lincoln's media strategy was successful. The telegraph-driven national media, which had first brought the lanky westerner to the country's consciousness, began reporting that Abraham Lincoln of Illinois was a serious contender for his party's presidential nomination. As news spread over the wires transmitting Lincoln's thoughts about slavery, the South was its tinderbox. When electric sparks delivered the news of Lincoln's election, the kindling ignited.

The telegraph technology that had helped magnify the crisis became a tool President Lincoln used to win the struggle. As the first national leader in history to use electronic communications for day-to-day governance and management, Abraham Lincoln became the first "online" president.[79]

When Lincoln arrived for his inauguration in 1861 there was not even a telegraph line in the War Department, much less the White House. When the U.S. Army wanted to send a telegram to a distant post they did as everyone else did and sent a clerk with the written message to stand in line at Washington's central telegraph office. The nation's leaders were, like their constituents, befuddled by electronic communications. Yes, newspapers, railroads, and financial and business institutions used the telegraph to move information rapidly, but just how the technology could aid the process of governing, much less in the midst of a war, was unknown.

Fourteen months into his presidency, Abraham Lincoln had his electronic awakening. Eerily, the awakening occurred eighteen years to the day from Samuel Morse's "What hath God wrought" message. When Confederate general Thomas Jonathan "Stonewall" Jackson changed the nature of the war by marching to threaten Washington, Lincoln responded by changing the nature of his leadership. Embracing the telegraph to bypass the chain of command and put some spark in his recalcitrant generals, Lincoln wired explicit commands to his generals in the vicinity of Jackson's advance. "You are instructed . . . to put twenty thousand men (20000) in motion at once for the Shenandoah," Lincoln ordered General Irvin McDowell. "Your object will be to capture the forces of Jackson & Ewell." Other orders to other generals followed.

It was an action unprecedented in the history of warfare. Never before had a national leader, based in the political capital, asserted himself to command troops in the field in real time. The telegraph office recently established in the War Department next to the White House became the nation's first Situation Room.

In the middle of a civil war, Abraham Lincoln harnessed a new network to make it an agent of his will. Lincoln's successful transformation of the mechanism of national leadership is even more remarkable because of its complete lack of precedent. No leader had ever used the technology as Lincoln did. Every leader thereafter would follow in his footsteps.

Although not specifically in reference to the new network, Lincoln's admonition in his 1862 message to Congress manifested the attitude that enabled him to see the opportunity in the telegraph and to seize it. "As our case is new, so must we think anew," Lincoln explained. His admonition is timeless. Driven by new networks, our case will always be new.

A Hint of Things to Come

The seeds of the information age were planted when the telegraph separated the speed of information transmission from the speed of transportation. The binary signaling that drives today's networks and devices began with the dots and dashes of the telegraph. Perhaps even more critical than its technological effects, however, is how the telegraph created the sociological and economic reality that drives our time. By removing time as a factor in the distribution of information, the telegraph introduced the modern imperative that if is it *possible* to possess information, then it is *necessary* to possess it.

Five years after Morse's fateful message, Patent Commissioner Thomas Ewbank's 1849 Report to Congress unknowingly wed the two technologies that would define the future. Writing about the wonderment of electronic distribution, Ewbank marveled, "Morse and his compeers have bridled the most subtle, fitful, and terrific of agents, taught it to wait . . . and when charged with a message, to assume the character of a courier whose speed rivals thought and approaches volition."

Then Ewbank saw into our generation: "If machinery don't *think*: it does that which nothing but severe and prolonged thinking can do, and it does it incomparably better." Channeling Babbage, the commissioner wrote: "In the composition of astronomical and nautical tables, accuracy is everything . . . [but] perfection in elaborate and difficult calculations is unattainable with certainty by human figuring . . . [thus] automata have been made to work out arithmetical problems with positive certainty and admirable expedition."[80]

It was the hint of things to come: the marriage of high-speed delivery of information with machines' ability to "think."

Connections

Alexander Graham Bell was working on improvements to telegraph technology when he discovered how to put sound on wires. The demand for telegraph service had precipitated a search for technologies that could compress multiple messages onto a single line and thus avoid the expense of stringing more wire. Bell had hypothesized that by sending telegraph signals at different musical pitches (frequencies) he could shoehorn several messages into a single wire simultaneously. To create his tones, Bell used the same technology as is used in a clarinet or other woodwind instrument, a vibrating reed. The dots and dashes created by the vibration of the reed would play out at the other end only when they resonated with a similarly tuned reed.

On June 2, 1875, one of Bell's reeds became stuck. When his assistant, Thomas Watson, hit the reed in an attempt to free it, Bell heard the resulting noise on the other end of the line.[81] Sound had been transmitted electronically! The same electromagnetism that made the telegraph sounder key rise and fall, Bell discovered, could carry sound when the sound-creating device on one end of the line had the same harmonic characteristics as the device on the other end.

That evening Watson constructed two devices in which an armature was mounted so that one end touched a stretched membrane and the other end an electromagnet. At the originating end a cone directed sound against the membrane so as to activate the electromagnet and produce a current. The other end, connected by wire, reversed the process as the armature, stimulated by the electromagnet, vibrated against the membrane. Watson recorded that he "could unmistakably hear the tones of [Bell's] voice and almost catch a word now and then."[82]

The electromagnetically controlled movement of an armature was driven by exactly the same physics that govern the telegraph. What Bell and Watson needed, however, was something more than the simple stopping and starting of a current flow. They needed to shape the current to resemble the acoustic signals of speech.[83] It took many months of being sequestered in the attic of Bell's Boston boardinghouse before the pair discovered the means of creating such a variable current.

To turn the sound into an electric current, Bell captured the sound waves on a membrane that vibrated against an armature connected to a battery. Then, to capture the variations a voice makes in conversation, impedance was introduced between the armature and completion of the circuit. The first successful impedance was provided by acid water. One end of a pinlike armature touched the membrane while the other end fluctuated up and down in the acid water and its electric current. As the needle vibrated, it moved closer to or farther away from the circuit contact immersed in the liquid. Though the acidic solution conducted electricity, it did so imperfectly. Thus the closer to the contact the needle came, the stronger the current it transmitted. Unlike the pulsing binary off and on of the telegraph, Bell's circuit produced a waveform similar to that of sound.[84]

On March 10, 1876, Bell spilled a bit of the acid and shouted for help. "Mr. Watson—Come here—I want to see you." As Bell recorded in his notebook that day: "To my delight he came and declared that he had heard and understood what I said."

"Mr. Watson—Come here" joined "What hath God wrought" in immortality. Later that evening, Alexander Graham Bell wrote to his father, "I feel that I have at last found the solution of a great problem and the day is coming when telegraph wires will be laid on to houses just like water and gas is, and friends will converse with each other without leaving home."[85]

In 1876, Bell filed for a patent on his discovery. The patent was described as "Improvements in Telegraphy."[86]

The telephone could be considered a technological step backward in the progression from the telegraph to the internet. Whereas the telegraph was a binary on-off signal, just like today's digital impulses, the telephone was an analog waveform—a speech-shaped electrical current. The telephone's contribution to the march to the digital future lay not in its technology but rather in the way in which it became ubiquitous. The technological step backward would become a monumental step forward when early digital information was disguised as a telephone call to piggyback on Bell's widespread network.

Part III

The Road to Revolution

The nineteenth century belonged to those who harnessed steam and sparks. By 1910, there were 351,767 miles of railroad track (compared with 204,000 miles of surfaced roads) linking the nation's villages and towns and drawing resources into urban production centers. As the tracks brought raw materials to central points to be processed, the people followed. The population of America's major cities increased seventeenfold between the early days of the railroad, in 1840, and 1910, by which time almost one-third of the nation's population resided in urban areas of greater than 100,000 people.[1]

Telegraph lines similarly spread like a skein across the landscape. By 1900, one company, Western Union, operated more than a million miles of wires carrying messages by spark.[2] Foretelling the future, Alexander Graham Bell's accident was also spreading. Shortly after the turn of the century, in 1907 there were 7.6 million telephones in the United States as the communications device that talked appeared in offices and homes.[3]

As these networks expanded into the twentieth century, they were incubating forces that would reshape that century and set the stage for the future. When these forces combined, they initiated the third great period of network revolution.

The networks of the nineteenth century had transformed the nature of physical connections by overcoming the constraints of distance and time. The networks of the twentieth century would add computational mathematics to interconnect virtually all information in incrementally costless delivery via a network of networks. Arriving at this point required a multistep process involving the computing devices themselves, the networks that connected them, and, ultimately, the delivery of connected computing power anywhere without wires.

The core breakthroughs that had thrice redefined physical networks became the foundation of the new virtual networks. The concepts in Charles Babbage's analytical engine evolved to become the computer. Binary electric signals that sent telegraph messages through regenerating relays evolved to become the logic circuits of those computers. And Johannes Gutenberg's disassembly of information into small pieces for subsequent reassembly became the format with which computers solved problems and networks exchanged information.

Five

Computing Engines

The cold December wind cut through central Iowa, slicing everything it touched, as the young professor took leave from the warmth of his family's evening dinner to return to his office on campus.

John Vincent Atanasoff, a thirty-four-year-old associate professor of physics at Iowa State College (now Iowa State University) in Ames, was obsessed with the idea that complex algebraic equations could be solved by a machine. "I went out to the office intending to spend the evening trying to resolve some of these questions," he would later recall, but "I was in such a mental state that no resolution was possible."[1]

It was 1937 and the state of the computational art had been unchanged for centuries; a machine could calculate (that is, add numbers), but human intervention was required to solve variable equations with a large number of calculations. The term "computer," in fact, referred to human beings sitting at endless rows of desks working with pencil and paper, slide rule, or mechanical calculator to produce one piece of a complex algorithm, which would then be combined with the work product of other "computers" in a laborious and slow march to an answer.

That cold December evening, Professor Atanasoff could find no inspiration at his desk. Thus, "I did something that I had done on such occasions. . . . I went out to my automobile, got in, and started driving."[2]

Leaving campus, he turned left onto Lincoln Highway and drove through Ames. Progressing eastward through Nevada (pronounced "Ne-vay-da" by the locals), the professor fell into a trance, with part of his brain minding the road ahead and the rest subconsciously churning away on his intractable problem.

Suddenly, he was almost 200 miles from Ames. The Mississippi River was approaching. It had been a surprisingly long drive that passed quickly owing as much to the professor's heavy foot as to the distraction of his musing. The opposite side of the Big Muddy beckoned with an opportunity not available in liquor-controlled Iowa: a roadside tavern and a glass of whiskey. "I drove into Illinois and turned off the highway into a little road, and went to a roadhouse, which had bright lights. . . . I sat down and ordered a drink. . . . As the delivery of the drink was made, I realized that I was no longer so nervous and my thoughts turned again to computing machines."[3]

At a corner table, away from the bar, sipping roadhouse bourbon and soda, John Atanasoff's thoughts began to sharpen. When he left the tavern, he had assembled in his mind the basic framework of a modern computer:

Electronics would replace the gears and levers of calculators to create a logic circuit.

Because electricity had two states—on and off—the machine would dispense with the decimal system's ten digits in favor of a binary digital system.

Vacuum tubes would provide the digital on-off signals; different on-off patterns among a collection of tubes would represent different numbers.

Memory storage would utilize capacitors capable of storing electricity that were "jogged" (Atanasoff's term) occasionally with additional electricity to keep them from losing what they were storing.

It was among the cornfields of Ames, Iowa, not at some highbrow eastern research university, that the world's first electronic digital computer came into being. During the winter of 1938–39, assisted by graduate student Clifford Berry, John Atanasoff assembled his ideas into a machine the size of a desk (76 inches long, 36 deep, and 40 high). Sitting in a basement corner of the Iowa State physics building, the Atanasoff-Berry Computer could solve twenty-nine linear equations with twenty-nine unknowns.[4]

The project's cost was $650, $450 of which was Berry's graduate stipend.

It was a discovery as seminal as Gutenberg's first tray of lead letters. Like Gutenberg, who saw his discovery suborned by financial chicanery, Atanasoff's breakthrough was soon pilfered.[5]

Calculating Machines

Until the first third of the twentieth century, the mechanization of mathematics had remained essentially unchanged for 5,000 years. The one exception was Charles Babbage, whom we last met in chapter 3 wistfully musing, "I wish to God these calculations had been executed by steam."[6]

Babbage's first attempt to mechanize math, the difference engine of 1822, was a mechanical iteration of the same concept the Babylonians had put to work with the abacus several millennia earlier. With its columns of tokens attached to rods representing units, tens, hundreds, thousands, and so on, the Babylonian abacus established the construct for mathematical manipulation: the adding or subtracting from each column to produce a conclusion.

The first effort to mechanize the functions of the abacus was in 1642.[7] Blaise Pascal, a nineteen-year-old French "wonder-jeune," created a shoebox-sized device, appropriately named the Pascaline, which contained interacting cogged-gear wheels. Windows on the top of the box displayed the numerical settings of the gears beneath. When the next number was dialed in, the gears acted just like the tokens on the abacus, moving the necessary number of notches and displaying the corresponding sum.[8] The nifty innovation of the Pascaline was a toggle lever between the gears that handled the "carry" function. When a gear finished its revolution the toggle would fall into place to advance the gear to the left by one position.[9]

Babbage's difference engine followed a similar construction of rods, gears, and levers, all powered by weights raised by a steam engine. Recall that Babbage was working on astronomical tables at the time of his exclamation about a steam-powered calculator. The calculation of these tables was a mind-numbing iterative process involving the repetitive application of common inputs (a multiplication table is an example of common input calculation). It was this constant application of a common input that started Babbage wondering whether the process might be susceptible to mechanization.[10]

Babbage built only a small proof-of-concept portion of the difference engine.[11] As he painstakingly laid out his concept in hundreds of pages of illustrations, however, by 1837 his thinking had broadened to a more advanced engine capable of using the results of a previous calculation to begin a new calculation.[12] Here was a conceptual breakthrough—a machine in which previous calculations would automatically feed it the next calculation. He called this concept his "analytical engine."

In his move from the difference engine to the analytical engine, Charles Babbage became the first person to close the gap between *calculating* and *computing*. A *calculating* device (whether an abacus, Pascaline, or difference engine) performs a single function on a single variable—adding or subtracting (or multiplying and dividing by repeated iteration). A *computing* device is capable of calculating multiple variables and then acting based upon its own computations. While the analytical engine still in many ways behaved like a giant Pascaline, Babbage had conceived of new functions that today we recognize as the essential components of a computer. He described them in nineteenth-century terms:

"Programs": instructions input via punch cards, a technique that would be used to input computer data well into the second half of the twentieth century.

"Store": the component to receive and hold instructions. Today we call it random access memory (RAM).

"Mill": the quaint industrial-age term applied to the component in which the numbers were manipulated by geared wheels interacting with each other. Today we would call it a central processing unit (CPU).

"Barrel": the control unit that instructed the mill when and how to operate. Babbage envisioned protrusions that pushed various rods to create the processing sequence, similar to the studded cylinder in a music box.

"Memory": to solve the carry problem, the adding function was separated from the carry both in function and in timing; the machine would hold a pending carriage while the addition was under way.

As with his previous effort, Babbage never physically constructed the analytical engine. Its intellectual breakthrough, however, survives in hundreds of pages of diagrams and descriptions.

At a time when the mechanization of mathematics was still based on the concepts of the abacus, Charles Babbage conceptualized a computing device.

The magnitude of Charles Babbage's breakthroughs defied the capabilities of the English language. Searching for a way to describe the concept of a machine that would produce a result that would guide its next action, Babbage fell back on the technological marvel of his age. The analytical engine was "a locomotive that lays down its own railway."[13]

Despite its status as one of the great intellectual achievements of the nineteenth century, Charles Babbage's vision of a computing device died with him in 1871.[14] Babbage's breathtaking concepts were relegated to the attic of quirky ideas.[15]

The Industrial Revolution was one of the culprits in the demise of Babbage's ideas. As industrial activity expanded, so did the need for the calculation of everything from boxcars to their contents, not to mention the large cash flows across multiple sources and uses. Science may need multivariable computation, but commerce required good old number crunching. It was in this period that the first "computers"—the human kind—were assembled to perform accounting functions, and increasingly complex calculators, derivatives of the Pascaline, began appearing on their desks.

Around the world, inventors took advantage of the industrial age's advances in machine tooling to make the gears of a calculator perform with precision similar to the gears in a Swiss watch. One such American inventor, a former bank clerk named William S. Burroughs, developed a manual calculator (all calculators were advanced by a manual function such as pulling a lever) in which the numbers were entered by pushing keys and the results were printed out on a roll of paper. The Burroughs Adding Machine Company became the dominant player in the calculator market.[16]

While Burroughs and others turned out mechanical calculators powered by cranks and levers, a former employee of the U.S. Census Bureau, Herman Hollerith, was marrying electronics to mechanics. The impetus was the magnitude of the growing nation's constitutionally mandated decennial census. The roughly 60 million returns from the census of 1880 took 1,500 clerks seven years to tally. There was fear the 1890 census wouldn't be completed before the next one was mandated to begin.[17]

Hollerith's solution, a punch-card-driven tabulator, was inspired by a Frenchman and a train trip. To protect against stolen tickets, train conductors of the time would punch each ticket to indicate the passenger's physical characteristics. "I was travelling in the West and I had a ticket with what I think was called a punch photograph," Hollerith recalled. By the strategic placement of punch holes on each passenger's ticket, "the conductor . . . punched out a description of the individual, as light hair, dark eyes, large nose, etc." Hollerith's aha moment was to record the census information on punch cards in the same manner.

Punch cards themselves were not a new concept. In 1801, the French weaver and merchant Joseph Marie Jacquard had invented an automatic weaving loom controlled by punch cards. As the punch cards moved through Jacquard's reader, wooden pegs fell into the holes to create the loom's patterns.[18] It revolutionized the textile business (and automated many people out of jobs). Babbage had envisioned punch cards as the input mechanism for the analytical engine.

Hollerith took Jacquard's process, applied it to numerical calculations, and electrified it. As the punch cards ran through Hollerith's machine, a collection of electrified pins would slide across the card; when a pin dropped through the punched hole, it fell into a vial of mercury that completed a circuit and sent a signal to the counter.

Six weeks after beginning the tabulation of the 1890 census, the Hollerith machines reported the population of the United States was 66,622,250. In addition to speed, the Hollerith machines offered the electromechanical ability to sort the information with unprecedented granularity. Pose any question, such as the number of farmers, or widows, or multigeneration families in a state or county, pull the appropriate cards for the target, line up the pins, and the numbers came to life.

Hollerith started a company to exploit his machine. The Tabulating Machine Company ultimately became International Business Machines, IBM.

From Calculating to Computing

While Hollerith's speedy tabulations were impressive, they weren't computing. To move in that direction—and to move back to where Charles Babbage was a century earlier—we need to meet two men on opposite sides of history's greatest conflagration.

Alan Turing was not the kind of person with whom you'd like to grab a beer. Rude, gruff, disheveled, and brilliant, he had few friends and even fewer intellectual equals.[19] In a 1936 paper, "On Computable Numbers," the twenty-five-year-old Englishman proposed that one of mathematical logic's great conundrums could be solved if only there were a machine capable of calculating algorithms. At a time when computation was done with paper and pencil, Alan Turing proposed something his contemporaries had never envisioned: a machine that automated algorithms.

Amazingly, Alan Turing had never heard of Charles Babbage, yet both conceptualized a computing machine—100 years apart. Babbage was obsessed with the mechanics of constructing such a machine. Turing could have cared less about building the machine; his was a purely intellectual exercise. It was an intellectual breakout of such magnitude, however, that the concept of an automated algorithm computer came to be dubbed a "Turing machine."[20]

On the other side of the English Channel, the rudiments of a Turing machine were coming together on a Berlin apartment floor. At the same time that Turing was intellectualizing in 1936, a twenty-six-year-old aircraft engineer named Konrad Zuse began building a Babbage-like universal calculator in the living room of his parents' apartment. Like Turing, Zuse had never heard of Charles Babbage.

The Zuse machine, like Babbage's, had a central processing unit where the computations were done, a control unit, memory, and a punch card/punch tape reader for inputting instructions. Unlike Babbage, however, Zuse substituted a base 2 binary numbering system in place of the traditional base 10 numbering. Zuse then used Boolean algebra (a form of algebra using only the binary symbols 0 and 1).[21]

The switch to a binary system simplified the mechanics. Whereas Babbage's device required a great number of complex interacting gears and levers, Zuse's binary calculator used simple slotted metal plates. A pin's posting in the slot on the left or right side determined whether it represented a zero or a one. In a fit of creativity, Konrad Zuse named his machine the Z1.

The Z1 had been a breakthrough in conceptualizing the operation of a computer. Zuse's next unit, the Z2, was a breakthrough in construction. In lieu of mechanical parts, the Z2 used secondhand telephone relays. Derived from Joseph Henry's 1831 telegraph relay, the technology that allowed telegraph signals to be reamplified became the source of on-off signals for a binary computing device. The on-off functioning of the relays made the Z2 like a self-contained telegraph network except that the on-off of the dots and dashes were now the 0s and 1s of binary code.

With World War II under way, Zuse's engineering skills were put to work in the German aircraft industry. In December 1941, he completed the Z3 to speed computational problems relative to the movement of aircraft wings under stress. Whereas the Z2 had been a prototype of a relay-based machine, the Z3 became the first operational general-purpose, program-controlled calculator.[22]

Konrad Zuse had built the Turing machine. All three Zuse machines were ultimately destroyed by Allied bombing raids.[23] Wartime secrecy, coupled with the Allied air supremacy that destroyed Zuse's work, meant Konrad Zuse's innovations had no effect on the path to electronic computing. Like Babbage, the world moved ahead ignorant of Konrad Zuse's discoveries.

And so we return to the cornfields of central Iowa. The trip to the Illinois roadhouse had taken place in December 1937; by August 1940 John Atanasoff's bourbon-based breakthrough was taking form in the basement of the Iowa State physics building. In December of that year Atanasoff traveled to Philadelphia for the annual meeting of the American Association for the Advancement of Science. There he met John Mauchly, the sole member of the physics department at nearby Ursinus College. Mauchly also had an interest in automating algorithms; he was developing an analog machine to trace the cycles of the weather. Atanasoff shared with Mauchly the basic concepts of his electronic computer. He even invited Mauchly to visit him in Ames to see the machine for himself.

A 1941 road trip halfway across the country was a major undertaking. Yet in June John Mauchly made it for the sole purpose of seeing John Atanasoff's computer. The visitor from the east moved into the Atanasoffs' home, saw the Atanasoff-Berry Computer (ABC) in operation, visited with its developer at length about its details, read the step-by-step thirty-five-page technical description of the breakthrough, and basically sucked out all the knowledge he could get from the Iowa professor. Then he returned to the East.

Less than three months after he left Ames and the ABC, John Mauchly wrote a paper in which he presented as *his idea* concepts similar to those Atanasoff had shared. Proving that science is also the domain of chutzpah, Mauchly continued to correspond with Atanasoff about the Iowan's breakthrough while conveniently failing to mention how his own work had usurped Atanasoff's ideas.

Then war came.

John Atanasoff left Iowa State in September 1942 for wartime duty at the Naval Ordnance Laboratory in Washington, D.C. The ABC stayed behind in Ames.

Back in Philadelphia, John Mauchly had joined the faculty at the Moore School of Engineering at the University of Pennsylvania. When the Moore School received a government contract for a machine to calculate ballistic tables for artillery, Mauchly became part of the team. The ideas purloined from John Atanasoff became the heart of the project.[24] Mauchly's machine, the ENIAC (for Electronic Numerical Integrator and Computer), would be heralded as the first electronic computer. He would leverage that claim to fame and fortune.[25]

ENIAC was a thirty-ton monster comprising almost 18,000 vacuum tubes and miles of wiring. It accomplished the previously unimaginable. The calculation for a ballistic trajectory that took humans twenty hours to solve took ENIAC thirty seconds![26] In fairness, John Mauchly and his collaborator, J. Presper Eckert, must be acknowledged for the manner in which they expanded upon Atanasoff's concepts and took them to scale. But the fact remains: they were Atanasoff's concepts.

While working on ENIAC, Mauchly and Eckert envisioned a new innovation. The ENIAC had been programmed by means of a plug board in which cables running between different inputs provided the necessary instructions. The next Moore School computer replaced the plug board with instructions stored in memory.[27] The Electronic Discrete Variable Arithmetic Computer (EDVAC) also finally broke with decimal mathematics in favor of binary numbers.[28]

When the war ended, Mauchly and Eckert left the Moore School in 1946 to start their own company. Like Hollerith, they saw opportunity in the calculations necessary for the national census. Their new company delivered the UNIVAC (UNIVersal Automatic Calculator) to the Census Bureau in 1951. It was the first commercial electronic computer. Interestingly, however, while the UNIVAC incorporated the innovation of storing commands in memory as the EDVAC did, it did not use the EDVAC's binary system but retained the base 10 decimal system.

The cash-flow realities of the new company ultimately led Mauchly and Eckert to sell the firm in 1950 to office equipment manufacturer Remington Rand.[29] Remington Rand merged with Sperry Gyroscope five years later to become Sperry Rand. In 1986 Sperry Rand merged with Burroughs Corporation (which had grown out of William S. Burroughs's calculator company) to become Unisys.

It was then that John Atanasoff got his due.

Mauchly's and Eckert's resignations from the Moore School in 1946 were preceded by a dispute with the school over whether they or the university would own the patent rights to the ENIAC concepts. The subsequent patents the duo filed were the underpinning of their commercial activities and the bane of competitors, who were obliged to pay royalties on the intellectual property. Things looked solid for the Mauchly-Eckert patents and Sperry Rand when in 1963 a U.S. district court judge ruled in their favor against a challenge to the patents. The judge specifically found an absence of evidence "of prior public use" of the ENIAC concepts.[30]

The royalty fees assessed by Sperry Rand on its competitors prompted one of them, Honeywell, Inc., to try again to invalidate the Mauchly-Eckert patent. This time the plaintiff had something new, the long-forgotten and never-heralded work of John Atanasoff, including how he had shared his discovery with John Mauchly. After nine and a half months of complex testimony U.S. District Court Judge Earl R. Larson ruled that "Eckert and Mauchly did not themselves first invent the automatic electronic digital computer, but instead derived that subject matter from one Dr. John Vincent Atanasoff."[31]

However, John Vincent Atanasoff's jinx on being recognized as the father of the electronic digital computer continued even as a court of law validated that fact. Judge Larson's ruling—surely worthy of national headlines proclaiming the real Father of the Computer—was handed down on October 19, 1973. The following day President Richard Nixon, knee-deep in the Watergate scandal, fired the Watergate special prosecutor, and the attorney general resigned in protest. The media, fixated on the "Saturday Night Massacre," overlooked the courtroom revelation about scientific intrigue at the dawn of the digital age.

Computing on a Thumbnail

Six years after the court decision, Mauchly's partner, Eckert, appeared in a Sperry Rand advertisement in the *Wall Street Journal*. Standing in front of Sperry's compact new 1100/60 computer, Eckert is juxtaposed with a photo of the huge ENIAC with the caption "Who would have thought that the father of Goliath would be introducing David?"[32] It epitomized the progress in computing machinery. Like the steam engines that first got Charles Babbage thinking about mechanizing calculations, computing engines were becoming ever smaller and ever more powerful.

The thirty-ton ENIAC would ultimately fit on a thumbnail. The first breakthrough on this path was the development of the transistor. It won Bell Labs scientist William Shockley the Nobel Prize in Physics.[33]

As we have seen, early computers used vacuum tubes as both an amplifier for electric signals and a switch. Shockley's transistor performed the functions of the vacuum tube but at a fraction of the weight, size, and power consumption. It was a trifecta made possible by a silicon semiconductor sandwich.

A semiconductor is a substance with an ability to carry an electronic current that lies somewhere between that of a good conductor and an effective insulator.[34] By adding impurities to a semiconductor such as silicon, Shockley could control its conductive characteristics. By sandwiching different types of silicon together he discovered it was possible to make electricity flow in one direction across the semiconductor. The word "transistor" was derived from *trans*ferring a current across a res*istor*.

Shockley left Bell Labs in 1955; moved to the Santa Clara Valley outside San Jose, California; and formed Shockley Semiconductors. Attracting bright young minds to join him and his Nobel laureate reputation, Shockley then proceeded to alienate them with his autocratic management style. In 1957 eight of his engineers—Jay Last, Julius Blank, Eugene Kleiner, Robert Noyce, Gordon Moore, Jean Hoerni, Sheldon Roberts, and Victor Grinich—announced they were leaving en masse. Even though Shockley had set the precedent, behaving similarly by leaving Bell Labs and taking his knowledge with him, he labeled them the "Traitorous Eight."

Soon, however, this group became the "Fairchild Eight." With investment support from Fairchild Camera and Instrument Corporation, they created Fairchild Semiconductor. There were now two semiconductor firms among the Santa Clara Valley's fruit farms. Silicon Valley was born. The eight built not only new technology but also a management model that featured stock options, little hierarchy, and a physically and functionally open work environment.[35]

Adapting semiconductor technology to produce a piece of electronic equipment in one solid piece, rather than multiple functions connected by soldered wires, became the Holy Grail of the research world. Early iterations by Texas Instruments soldered wires so small they required tweezers and a microscope to place onto a circuit. It was the same concept as a typical circuit board, but rather than wiring various components on a board, these components were all on one solid circuit semiconductor. It was called an integrated circuit.[36] At this early stage, however, the fragile soldered connections affected the solution's practicability.

Two years after the founding of Fairchild Semiconductor, in January 1959, Robert Noyce invented the first practicable integrated circuit. By etching semiconductor circuits on a piece of silicon, Noyce created microchips that could quickly and efficiently move electric circuits. Others on the team developed and built the processes and equipment necessary to produce Noyce's development.[37]

All during these technological breakthroughs, Noyce and Gordon Moore were unsuccessfully trying to convince their parent corporation to embrace a new management philosophy that would make stock options available to a broad group of employees. While successful in their technological innovation, they did not succeed in finding favor for their management innovation at Fairchild headquarters. Once again Noyce and Moore took what was between their ears and departed (other members of the Fairchild Eight had pulled the ripcord earlier). It was a gutsy move; the company in which they had invested a decade of their lives had seen its revenues grow from nothing to $130 million by 1967.

Noyce and Moore's new outfit opened for business in 1968. They named their new company Intel.

The business plan for Intel was to produce integrated circuits that could be used as memory for computers. This was an ambitious goal, as at the time such technology was at least a hundred times more expensive than storing data on magnetic tapes.[38] In 1970, however, Intel invented the dynamic random-access memory chip (DRAM) that reduced those storage costs.

When Japanese calculator manufacturer Busicom approached Intel about fabricating a dozen different chips, each with a separate function, that could be strung together for use in a programmable desktop calculator, a young Intel engineer, thirty-one-year-old Marcian (Ted) Hoff, saw an opening for a new kind of integrated circuit. Instead of purpose-built logic chips, Hoff envisioned a programmable chip that could act like a conventional central processing unit. The client liked the idea, and the Intel 4004 was developed, a programmable computer on a single chip—a microprocessor.

Such an idea did not harmonize with Intel's business plan, however. The company produced memory chips, and Hoff's idea was a far cry from those (profitable!) products. To make matters even more challenging, Intel did not own the microprocessor technology Hoff had invented. Because the work had been done under contract to Busicom, the Japanese firm owned it. Busicom, however, had fallen into difficult financial straits. In order to raise cash, it was willing to sell the technology back to Intel for $60,000.[39]

Other than Busicom's desk calculators, however, there was no established market for microprocessors. Would Intel risk $60,000 of its limited capital to buy back a technology with no current market and no place in its business plan? More important, could Intel grow two breakthrough technologies simultaneously, and how would the company's pursuit of the DRAM market change if it redirected resources to try and build a microprocessor market? Except for Moore and Noyce, the senior managers were against the move into microprocessors.

Gordon Moore and Robert Noyce, however, championed the microprocessor idea in a creative way. Since every microprocessor would also require DRAM, they argued that it was a strategy to sell more memory.[40] The pitch prevailed—it was like investing in the steam locomotive as a way to sell coal! It was "literally betting the company," an Intel executive would tell me years later.[41]

In 1971 Intel brought the Intel 4004 to market.[42] Intel's first microprocessor had as much processing power as ENIAC-type machines that sold for hundreds of thousands of dollars.[43] Yet the Intel chip was priced at hundreds of dollars. Four months later Hoff and colleagues produced the more powerful Intel 8008, and within a year they were producing the Intel 8080, which ran 220 times faster than the Intel 4004.[44]

Big computers, however, were still the computing mainstay. By the mid-1970s computers had become like the hand-copied books pre-Gutenberg—sacred knowledge and capabilities locked away in facilities inaccessible to mere mortals, where they were tended by a special priesthood. Microprocessors may have put the Turing machine on a chip, but the priesthood could not fathom its relevance to everyday life.

The kind of people who read *Popular Electronics* magazine saw the opportunity differently, however. In 1973, the magazine asked its readers to contribute their design for "the first desktop computer kit." The winning entry from Edward Roberts in Albuquerque became the cover story of the magazine's January 1975 issue. The kit was powered by an Intel 8800 microprocessor. Roberts's daughter suggested he name it after the planet on the *Star Trek* episode they had just watched—Altair. The kit cost $397. It had no keyboard, monitor, or operating system. Data were loaded via toggle switches, and the results of the computer's efforts were displayed on the front panel. Roberts received 4,000 orders in the three months following the cover story. The personal computer was born.[45]

Two readers of the January 1975 issue of *Popular Electronics* were a Harvard student named Bill Gates and his friend Paul Allen. Paul Allen called Edward Roberts and told him the two had programming software for the Altair. They called their company Micro-Soft.[46]

Two years later, two Silicon Valley whiz kids brought to market a $790 fully assembled personal computer. Steve Wozniak's and Steve Jobs's Apple II moved personal computing beyond kit-building hobbyists. Software programs written by others allowed the Apple II to do word processing, create spreadsheets, and play games. By 1981, Apple had $300 million in annual sales and employed 1,500 people.[47]

In 1981, the priesthood came after the upstarts. Mainframe powerhouse IBM introduced the IBM PC and chose Intel for the PC's microprocessor. IBM turned to Bill Gates and Paul Allen for the computer's operating system. Combining the initials "IBM" and "PC" seemed a contradiction in terms, but it gave personal computing the credibility boost that changed its course. In 1981 IBM sold about 35,000 IBM PCs; two years later sales of the microprocessor-powered devices had soared to 800,000.[48]

Even with that kind of sales growth, PCs were still foreign to most people. Today we cannot imagine our lives without microprocessors and PCs, but it was only a short time ago that individualized computing was a concept from science fiction. In 1984, for instance, I was CEO of a computer and networking company; our salespeople's greatest challenge was to get prospective customers to *touch* the computer or its keyboard.[49] So ingrained was the idea that computers were mystical and fragile that we trained our salesforce to seat the customer and then physically place the keyboard in the customer's lap to force them to touch it.

The trip from Charles Babbage to the personal computer was a century-and-a-half odyssey. It was, however, the first step in defining our current network revolution. Soon the ideas of Charles Babbage, Alan Turing, Konrad Zuse, and John Vincent Atanasoff became manifest in our daily lives through the ability of computers to communicate with one another and with networks of computers.

Connections

For the thirty-fifth anniversary issue of *Electronics* magazine in 1965, Intel cofounder Gordon Moore (who was still at Fairchild at the time) wrote an article titled "Cramming More Components onto Integrated Circuits." The article made the seemingly wild forecast that the number of transistors on an integrated circuit would double every year, bringing processing power up and cost down. The result of this, Moore predicted, would bring "such wonders as home computers . . . automatic controls for automobiles, and personal portable communications equipment."[50]

Dubbed "Moore's law," that 1965 forecast has proven incredibly resilient over the past fifty years. While the complexity of chip making has caused the time component of Moore's law to vary, its basic concept still holds true: every couple of years the computing power of the state-of-the-art microprocessor doubles exponentially. When Gordon Moore wrote the article, the most prominent challenge was the shift from thirty transistors to sixty on a microchip; today, the number is in the billions.

At age fifty, Moore's law is beginning to show its years and slow down, but its trajectory remains on course. In the process it has become the computational bedrock of the new network revolution. The exponential functioning of Moore's law means that today's smart phone has the computing power of the multimillion-dollar supercomputer of just a few decades ago. Of even greater significance to our future, however, is the continued up-and-to-the-right growth of microchip computing power, with the effect that the computing power increases we will see in the next five years will far exceed what we experienced in the last five.

Six

Connected Computing

A vibrant palette of raucous colors and crisp air welcomed the members of the American Mathematical Society to Dartmouth College in Hanover, New Hampshire, for their 1940 conclave.

Standing before the audience gathered in McNutt Hall on September 11, George Stibitz delivered a paper describing the Complex Number Computer he and his colleagues had constructed 250 miles away at Bell Laboratories in New York City.[1] At the heart of Stibitz's computer was a discovery he had made on his kitchen table. He had repurposed the on-off functionality of the electromagnetic relays typically used to route telephone calls to instead solve mathematical problems.[2]

The discovery Stibitz was demonstrating was developed on a November weekend in 1937. Using a couple of telephone relays, tin strips cut from a tobacco can, a flashlight battery, and some flashlight bulbs, Stibitz created a contraption that added binary numbers to produce a binary sum. In honor of the kitchen table at which it was invented, Stibitz dubbed it the "Model K."

Less than three years later he stood before his peers describing how he had harnessed the on-off functionality of approximately 400 telephone relays into a machine that could add, subtract, multiply, and divide complex equations.[3]

Then George Stibitz turned showman.

At Bell Labs, Stibitz's Complex Number Computer was locked away in a small room. Modified teletype machines outside the room provided inputs to the machine. Operators would type the problem to be computed into the teletype, the impulses would be fed to the computer, and the teletype would print the result. Stibitz had arranged for a similar teletype to be on stage with him at Dartmouth. He would not only describe his computer; he would make it perform for the crowd as if it were a circus animal, even though it was 250 miles away.

It was the first time a computing machine had ever been accessed remotely via phone lines.[4] On stage, Stiblitz would input a problem into the teletype terminal on stage; the problem would be transmitted 250 miles to the computer in Manhattan, where the computation would be performed and sent back to the awaiting teletype. The whole 500-mile round trip and processing took about a minute. It seemed like magic.

Like a carnival barker, Stibitz invited the audience to come forward and try his computer. From 11:00 a.m. to 2:00 p.m. that day attendees played Stump the Box. The box always won.

The cause of the excitement, as well as the topic of Stibitz's paper, was the Complex Number Computer and its capabilities. The history that was made that day, however, was the remote access of a computing device over telephone wires.[5] The technology Alexander Graham Bell had developed would become the infrastructure for communicating computers.

A Technological Step Backward

As we've seen, the telephone could be considered a technological step backward in the progression from the telegraph to the internet. While the telegraph was a binary on-off signal just like today's digital impulses, the telephone was an analog waveform. The telephone's contribution to the march to the digital future was not its technology alone, but how that technology developed into a ubiquitous backbone for connecting computers.

The path to George Stibitz's demonstration began in 1871 when the Boston Board of Education recruited a young Canadian who was fascinated with acoustics and the nature of sound to teach hearing-impaired students. The teacher's name was Alexander Graham Bell.

The dominant network company at the time, Western Union, was in search of a technology that would increase its productivity and lower costs by allowing a single wire to carry multiple messages simultaneously. Thomas Edison, while on retainer to Western Union, had invented a means for four signals to be carried simultaneously—two in each direction.

Edison—a man whom Western Union president William Orton described as having a "vacuum" where his "conscience ought to be"—contended he owned the patent on this quadruplex technology despite the fact he had developed it under contract to Western Union.[6] Financier Jay Gould, who was engaged in manipulating Western Union's stock, bought Edison's patent. By threatening to employ the patent in his rival telegraph companies, Gould was able to depress the market value of Western Union.[7]

Riches lay ahead for anyone who could replicate Edison's quadruplex effect without violating his patent. Back in Boston, the father of one of Bell's students, prominent attorney Gardiner G. Hubbard, financed the acoustics expert so he could apply his knowledge in pursuit of harmonic transmission whereby multiple different signals would move on the same wire at different frequencies.[8]

There was also another dynamic at play for Bell and Hubbard; Bell and his student Mabel Hubbard were in love. Arguably, one reason to accept Hubbard's commission was for Bell to curry favor with his future father-in-law.[9]

Bell conceived of what he called an "undulatory current" to carry the telegraph signal. Presumably, multiple such signals operating at different frequencies could carry multiple telegraph signals on a single wire. Bell's acoustic training also convinced him such an undulatory current could carry sound. "If I can make a deaf-mute talk," he reportedly said, "I can make iron talk."[10] On February 14, 1876, Alexander Graham Bell filed his patent on undulating signals.[11] A few weeks later, on March 10, the historic "Mr. Watson—Come here" transmission moved Bell's voice over wire.[12]

At first the ability to "make iron talk" was more of a phenomenon than a practical application. At the Philadelphia Centennial Exposition of 1876, Bell's demonstration was banished to a small table in an out-of-the-way corner. For six weeks it sat there unappreciated until Hubbard (who was one of the commissioners of the Centennial Exposition) persuaded the judges to detour and take a look. As if by providence, as the bored judges were at Bell's display (reportedly not even putting the receiver to their ears), the emperor of Brazil swooped in and greeted Bell by name (he had visited one of Bell's classes for the deaf in Boston). His warm welcome got everyone's attention. Picking up the receiver and listening, the emperor exclaimed, "My God—it talks!"

Soon the venerable Joseph Henry, the very man whose ideas Samuel Morse had purloined for his own, proclaimed, "This comes nearer to overthrowing the doctrine of the conservation of energy than anything I ever saw." Sir William Thompson (later Lord Kelvin), the world's foremost expert on electrical energy, proclaimed, "It DOES speak. . . . It is the most wonderful thing I have seen in America."[13] Suddenly, Bell's invention was lifted out of obscurity and onto the exposition's central stage.

The Bell phenomenon went on a road show after the exposition. Transporting voices and music into halls throughout New England, the new technology, while greeted with awe, bumped up against the same resistance prompted by earlier network innovations—including the accusation that it had to be the work of Satan. In an echo of Baltimore's pastors warning of Morse's telegraph, the *Providence Press* observed, "It is hard to resist the notion that the powers of darkness are somehow in league with it." The *New York Herald* described it as "weird and almost supernatural."[14]

It was a long way from road-show demonstrations in public halls to a universally available network that switches calls between users. The new technology was so revolutionary that there were no readily apparent applications for it. To Bell's financier, Gardiner Hubbard, fell the burden of seizing the public's interest to turn the technology into a business. One by one, pairs of telephones slowly found their way into use, but there was still no tsunami of demand. An early application was to call (or receive a call from) the central telegraph office to send (or receive) a telegram, but the telephone remained a technology in search of a market.[15]

With business dragging and capital hard to raise, Gardiner Hubbard acquiesced to the inevitable. In the winter of 1876–77, he offered Bell's patent to Western Union for $100,000. William Orton, the president of Western Union, rejected the proposal, reportedly (perhaps apocryphally) with a sniff: "What use could this company make of an electrical toy?"[16]

Orton's rejection of Bell's patent made the incumbent networks zero for two in recognizing the opportunity of a new network technology. In 1845, the Post Office Department had turned down Morse's patent and its $100,000 price tag. Thirty-two years later, Morse's progeny made the same short-sighted mistake for the same $100,000 figure.

Enter a New Vail

Alfred Vail, Samuel Morse's underappreciated assistant, played a pivotal role in the development of the telegraph. His younger cousin, Theodore Vail, would be seminal in the development of universal telephone service.

At thirty-three, Theodore Vail had already distinguished himself as a hard-charging Post Office employee who had improved the efficiency of the railroad mail service and ultimately had 3,500 people reporting to him.[17] In 1878 Gardiner Hubbard recruited Vail to run the year-old Bell Telephone Company.[18] When Vail tendered his resignation, the assistant postmaster general, his boss, incredulously responded, "I can scarcely believe that a man of your sound judgment . . . should throw it up for a d____d old

Yankee notion . . . called a telephone!"[19] It was not an illegitimate observation. At the time the Bell Telephone Company had all of 10,000 phones in service and the first manual switchboard had just been introduced.[20]

His new job at Bell confronted Theodore Vail not only with building a market for telephone service and all the technical issues associated with network expansion but also with the challenge of executing basic business activities, including raising capital. Most threatening to the small company's future was the newly awakened Western Union. By 1878, Western Union had reversed its position of two years earlier and entered the telephone business. Thomas Edison, his relationship with the company restored, had significantly improved Bell's telephone devices so that Western Union subscribers did not have to shout for their devices to work. By the end of 1878, Western Union's American Speaking Telephone Company subsidiary had 56,000 telephones in service.[21]

Alexander Graham Bell became so depressed at how Western Union had assumed dominance over the technology he had invented that he checked himself into Massachusetts General Hospital.[22]

Western Union's strategy was multifaceted. Backed by the deep pockets of the nation's largest corporation, Western Union lawyers descended on the Bell patent to challenge its validity. Reaching into those deep pockets again, Western Union would buy out local Bell franchises. Where the Bell licensees would not sell, Western Union overbuilt them with a competitive offering.

Alone on the Bell battlements stood Theodore Vail and his small band.[23]

Amazingly, Western Union blinked. Some attribute this to the company's ongoing struggle with the financier Jay Gould which served to challenge management.[24] Others assert the Western Union lawyers came to the realization that theirs was the weaker patent position.[25] Whatever the reason, the behemoth parleyed with the upstart. The result was a market-dividing pact. Western Union would concede the Bell patent and depart the telephone business. In return, the Bell Company would purchase Western Union's telephone assets, pay a royalty to Western Union on each piece of telephone equipment, and agree not to enter the telegraph business.[26]

With a clear path ahead, Theodore Vail pushed to expand telephone service out of local municipalities by interconnecting local exchanges. "We have a proposition on foot to connect the different cities . . . to organize a grand telephone system," Vail wrote in 1879.[27] Vail's vision was derided as unrealistic. Placing calls within a city had been a slow-growing activity—why was there a need to call between cities?

It fell to Vail to personally organize a new company outside Bell for the purpose of building a line between Boston and Providence, Rhode Island. Such foolishness was dubbed "Vail's Folly." Today the success of the Boston-Providence line should come as no surprise. At the time, however, it was dumbfounding. Yet thanks to Vail's intercity concept, the telephone became more than a neighborhood affair.[28] The following year, a new line from Boston to New York went into operation.

The bridge to the universally connected network was under construction.

A dispute with Bell Company directors over whether to reinvest profits into the network or pay dividends drove Theodore Vail to take his leave of the company in 1887.[29] Having bested Western Union and battled to interconnect local telephone exchanges, Vail threw in his hand and sailed for entrepreneurial adventure in South America.

Vail's Vision Triumphant

Vail's new venture built the company that provided electricity for the lighting and trolleys of Buenos Aires. When Vail left the United States in 1887, he was well-off; when he returned barely more than a dozen years later, he was truly wealthy. Rich and out of the day-to-day combat of commerce, Vail became a gentleman farmer on his Vermont estate.

During his absence the telephone landscape of the United States had changed dramatically. Bell's patent had expired in 1894, and independent telephone companies were springing up everywhere. By the time Vail returned to enjoy his riches, the independents' 3 million users exceeded the Bell System's 2.5 million. Strategically, the independents, which controlled most of the West and the nation's rural areas, had the wherewithal to threaten Bell's dominance in the major cities. The Wall Street bankers who had wrested control from the Bostonians who had overseen the Bell Company since the days of Alexander Graham Bell persuaded Vail to return to the company (now renamed American Telephone & Telegraph) in 1902 as a member of the board of directors. Five years later, in May 1907, a delegation of fellow directors appeared at his Vermont estate to ask Vail to return to the helm of the Bell System.

Twenty years after a dispute over corporate vision forced Theodore Vail out of Bell, he was back as the boss.[30] He was sixty-two. More than any other individual, Theodore Vail is responsible for the existence of the ubiquitous national network that was necessary for the launch of the information age. At the same time, Vail also bears responsibility for policies that stymied the ultimate rollout of digital service.

As the new president of AT&T, Vail imposed policies that had been rejected twenty years earlier. At the heart of his vision was the concept of "one policy, one system, one universal service." Vail set about to make his company the sole provider of that three-pronged offering.

In the company's annual report for the year of his return, Vail laid out his philosophy. "The strength of the Bell System lies in its universality," he explained. Then he declared war on competition. "Two exchange systems in the same community, each serving the same members, cannot be conceived of as a permancy [*sic*], nor can service in either be furnished at any material reduction because of the competition, if return on investment and proper maintenance be taken into account. Duplication of charges is a waste to the user."[31]

A quarter century earlier Theodore Vail had been a competitive crusader fighting to keep Western Union from expanding its market dominance. At the helm of AT&T, he became the principal advocate for the concept of a "natural monopoly"—the notion that economic efficiency would be undermined by the existence of multiple firms and enhanced by the scale achievable by a single firm. It was the pre-antitrust era of integrated market dominance throughout the economy. Theodore Vail intended to become the integrator and dominant force in the electronic communications business.

Vail immediately began buying the competition. The Bell System swallowed independent telephone companies. Those who chose continued independence found it difficult to connect with AT&T's long-distance network. When independents were allowed to connect, the Bell System dictated their operations as a condition of that interconnection.

Then Vail bought his old nemesis, Western Union. In 1879, Vail had warned of the effect of a monopoly controlling both telegraph and telephone. By 1909 he was that monopoly.

To achieve his "one policy, one system, one universal service" vision, Theodore Vail enlisted an unlikely partner: the government. As concern about monopolies—even "natural monopolies"—expanded, Vail made government both a watchdog and a co-conspirator. "Private management and ownership, subordinate to public interest and under rational control and regulation by national, state, and municipal bodies is the best possible system," he expounded.[32] In 1913, Vail agreed with the federal government to divest the newly acquired Western Union, cease from acquiring competitors, and to interconnect his long-distance services with other telephone companies to provide long-distance service. This agreement, named for Nathan Kingsbury, the AT&T general counsel who negotiated it (the "Kingsbury Commitment"), convinced the government to back off its antitrust inquiry into the company's activities. It was the embodiment of Vail's vision of a symbiotic relationship between his monopoly and government. There would be only one provider of telephone service per market and only one long-distance company to connect them. With the imprimatur of the government in return for regulatory oversight, the Vail vision was complete. The company flourished as service expanded.

The Vail-built troika of universal service, based on a government-regulated monopoly and driven by massive revenue with small margins, would define telecommunications for the twentieth century.

Competition and Innovation

In his 1910 report to stockholders Theodore Vail expounded on his vision: a "universal wire system" for the "electronic transmission of intelligence."[33] He was prescient in the synergy between universality and the electronic transmission of intelligence; the analog phone network would become the universal gateway to early digital activity. He was less insightful about the impediment to such innovation that his monopoly created.

Vail's success in transferring AT&T's focus away from a competitive marketplace had the effect of creating a sclerosis that hardened the network against change. While competition traditionally drives innovation, the company Vail created had succeeded in eliminating competition and thus could innovate (or not) at its own pace. The universality of the network Vail created may have become the threshold enabler of the information age. Yet that innovation would not occur without a fight, as the monopoly's antibodies fought against the threat of disruptive innovation.

Under Vail, innovation was to be embraced but controlled. Vail, who had watched as Jay Gould used Edison's technology to manipulate the activities of Western Union, appreciated how new technology could negatively affect an established business. He saw a means of mitigating such a threat if the company became a technology leader itself. He began a process of expansive technology development that looked into the needs of the future, but on Bell's terms.[34]

In 1925, AT&T created Bell Laboratories, which ultimately became home to arguably more IQ per square foot than any other place on the planet. Bell Labs engineers originated ideas as important as Claude Shannon's in "A Mathematical Theory of Communication," which envisioned information as a physical quantity to be manipulated. *Scientific American* called it "the Magna Carta of the information age."[35] Bell Labs was the progenitor of the essential components to fulfill Shannon's thesis, including the transistor, magnetic storage, and early computer languages.

AT&T became a frenetic propagator of innovation, but a sluggish adopter. While its leaders understood that the nature of networks must ultimately evolve, the first purpose of the corporation was to preserve the market position of the present network. One example of this policy at work was the development of magnetic storage. In the early 1930s a Bell Labs engineer, Clarence Hickman, developed the first answering machine by developing magnetic tape that would record sounds. It was a discovery about the magnetic storage of information with implications far beyond recording a missed phone message. AT&T ordered Bell Labs to cancel Hickman's research because management worried that if the public had the ability to leave a message people would place fewer phone calls.[36]

The ultimate tool to exploit the symbiotic relationship between the government and the phone company was a federal rule (actually an FCC-approved tariff describing rates and services) that gave the AT&T monopoly total control over anything that attached to the network. The company used the full force of the federal government to prohibit the use of any "equipment, apparatus, circuit or device not furnished by the telephone company."[37] In a fit of public relations inspiration, such non-Bell devices were labeled "foreign attachments," or sometimes "alien attachments." The specter that a non-Bell "alien" device could cause physical harm to the network, perhaps even disabling the national defense system, was a carefully cultivated myth.[38]

The telephone company may have successfully built a universal pathway, but its absolute control of that pathway meant that AT&T, and through it the United States, would enter the digital age when AT&T damn well pleased. An example of this chilling control was the digital modem, the tool that ultimately opened the door to the internet.

Whereas the telephone delivered sounds via a smoothly undulating signal (called a waveform), a digital signal is choppy with square, discrete components. In order to carry a computer's digital output over an analog phone line it is necessary to remodulate the digital signal into sounds that made it similar to a voice signal. The term "modem" is shorthand for how a device modulates digital information into analog signals (the screeching you hear when two fax machines connect) for transmission as if it were the sounds of a voice, and then demodulates it at the other end into digital format for the computer.

Since a modem connected to the telephone network, AT&T controlled both the unit's capabilities and its availability. Using the power to dictate standards for connecting to the network—a ploy that Theodore Vail first used to impose his will on independent telephone companies—AT&T dictated the availability of modems, their cost, and their design. When the federal courts finally saw through the charade and began to chip away at the absurdity of the "foreign attachment" claims, the door to innovation in modem development and design began to open.[39]

Gradually, as the rules that allowed AT&T to unilaterally determine who and what attached to the network disappeared, the nation began to utilize the universality of the telephone network to deliver computer-to-computer information on a widespread, generally available basis. In the early 1980s, following the government's decision to eliminate the restrictive rules, a new generation of inexpensive non-Bell modems came on to the market and the information age became manifest for anyone with a personal computer and a phone line.

From Data Transmission to Digital Networks

Our story of network evolution now requires a look back to the birth of digital packet switching discussed in chapter 1. Paul Baran's 1964 paper, "On Distributed Communications," identified how the hub-and-spoke analog telephone network could be made less vulnerable to attack by adopting a fishnet-like topology and digital packet routing.[40] In the process he had also identified a potentially more efficient, lower-cost network technology than that used by AT&T.[41]

When the Department of Defense went to AT&T to request that the phone company adopt Baran's ideas, it was turned down cold. "The Air Force said to AT&T, 'Look, we'll give you the money. Just do it,'" Baran would later recall. "AT&T replied, 'It's not going to work. And furthermore, we're not going into competition with ourselves.'"[42]

The "it won't work" response is not illogical. Baran's concept meant trashing network truths that had been drilled into the brains of generations of engineers. The advocates for change described these monodimensional individuals as having "Bell-shaped heads."

Ever since "Mr. Watson—Come here," communications networks had been about establishing an open circuit between two points and maintaining that circuit for the duration of the transmission. It was a highly inefficient technology for which circuit capacity had to be built in anticipation of peak demand. Such costly inefficiency, however, was the economic backbone of Theodore Vail's "natural monopoly."

The connection between network economic inefficiency and monopoly underpinned Baran's other observation about the phone company's unwillingness to embrace a new, potentially competitive technology. The executives entrusted with Theodore Vail's monument, the world's greatest communications network and the mainstay of the stock holdings of widows and orphans, were not about to change their business model and the guarantee of a fixed rate of return on every dollar invested, even if that meant using inefficient infrastructure.

Thus, when Paul Baran presented AT&T with a network concept that increased efficiency by not holding a circuit open, but rather opening and closing it multiple times a second in order to load it chockablock with digital packets, the Bell System executives rolled their eyes and ran the other way. "I might as well have been speaking Swahili," Baran once told me.[43]

Contrary to urban legend, the internet was not built as a means of surviving a Soviet attack. Such survivability had been the impetus for Paul Baran's work that resulted in the packet-switching technology now used in the internet, but Baran's second-strike network was never built. In fact, after the responsibility for such a network was transferred to the Defense Communications Agency (DCA), Baran worked actively to scuttle the project. "I told my friends in the Pentagon to abort this entire program—because they wouldn't get it right," Baran recalled. Having just fought circuit engineers at AT&T, Baran wasn't anxious to leave his baby in the hands of yet another group of analog-minded engineers at DCA who, because they didn't get it, would condemn the concept to failure. "The DCA would screw it up and then no one else would be allowed to try, given the failed attempt on the books," Baran explained. He would rather wait until "a competent organization came along."[44]

The Defense Department did build their own packet-switched network—but it was not for controlling bombers and missiles. In 1969 the department's Advanced Research Project Agency (ARPA) connected the networks serving the mainframe computers of four U.S. research institutions so that scientists could share data and access each other's computers.[45] The project was known as ARPANET.[46]

It seems such a simple concept today, but ARPANET was a huge reach, both technically and philosophically. While not the internet, ARPANET was the internet's opening act. Research institutions had acquired large mainframe computers, each accessible only via the institution's private network. As in the early days of the telephone, these were purely local networks. But what would be the effect if they could be interconnected, just as "Vail's Folly" had connected Boston and Providence?

ARPANET was not only a technological feat in connecting private networks to create a giant computer time-sharing network. Perhaps even more daunting, it required a cultural change as well. Competitive institutions signed up knowing full well that they were joining an "open" network that allowed access to each other's previously sacrosanct computers.[47]

ARPANET was a hit in the academic world. Two years after its launch connecting four supercomputers, twenty-three universities and government research facilities were connected. By 1984 there were more than a thousand ARPANET hosts.[48] The connections between the ARPANET nodes were Theodore Vail's universal network.

Other data networks were starting up as well. The functionality of packet switching was adopted by commercial network service providers such as Tymnet, Telenet, and CompuServe. Essentially, each of these (and other) networks was using the efficiency of packet switching inside the network, but with its own particular protocols. This made the networks incompatible with each other. Information on one network had to be transformed in order to work on another. Theodore Vail had solved this problem by bludgeoning independent phone networks to adopt AT&T's standards as a condition of interconnection; now it was necessary to develop similar standards, absent the bludgeon. In 1972 the renamed Defense Advanced Research Projects Agency (DARPA) commissioned a project to tackle this problem. It was called the "Internetting Project" and the term "internet"—a network of interoperating networks—was born.[49]

At the heart of this internet was a lingua franca known as the TCP/IP protocol suite. For their work in developing this lingua franca, Robert Kahn and Vinton Cerf have appropriately been dubbed the "Fathers of the Internet."[50] On New Year's Day 1983, all of the host computers on ARPANET adopted the TCP/IP format. It remains today the rules of the road for the internet.[51]

The first challenge Kahn and Cerf faced was to define a uniform addressing methodology. "Packetizing" the information into small unique units of data begins the process. Then Internet Protocol (IP)—the principal communications protocol for the internet—delivers the digital packet to the right address. As shown in the figure in chapter 1, the digital-packet-switched network looks like a fishnet. At each intersection of this labyrinth is a small computing device known as a router. These routers chat back and forth with nearby routers to learn about each other—whom they connect to, and their availability and status. When a packet arrives, the router looks at the database created by this gossiping for the IP address of its destination to pass the packet off to another router in the general direction of its destination. All of this occurs in a fraction of a second.

But with each packet traveling independently, the packets don't arrive in the same order at their destination as they were launched. This is when the Transmission Control Protocol (TCP) goes to work. Each packet also contains information about how it fits into the whole of the collected packets. At the destination, TCP reassembles the packets in the correct order.

Many mark the birth of the internet with the implementation of TCP/IP protocols. While ARPANET previously had been about *connecting* computers, TCP/IP made it one giant network of *interacting* computers. The "internetworking" that allowed a collection of disparate but connected networks to interoperate through the same language became the internet's hallmark.

Concurrently, the other determining characteristic of the internet was also in development. As the number of routers in the network expanded, the gossiping between them became inefficient. In the early 1990s, a new Border Gateway Protocol moved that routing function further to the edge of the network. With routing now accomplished at the borders of the network, the fishnet topology that Paul Baran envisioned in 1964 had become a reality.

But these developments were all about computers *connecting* with each other. What about *finding* information on these networks? And even if you did find what you were looking for, would it be in a format that was usable? Those issues were solved in 1990 when Tim Berners-Lee developed a means to identify, retrieve, and interrelate networked information.

Berners-Lee called his development the World Wide Web. When the internet became the web, it leapt out of the world of computer science into everyday usability and functionality.

While the terms "internet" and "web" have become synonymous, they function very differently. The web identifies and retrieves information using the server-client architecture enabled by the internet.

At its heart, the web consists of three components. First is a unique identifier for the information, the Uniform Record Locator (URL) that we know as the "web address." The common language to publish that record is "hypertext markup language" (HTML). And the "hypertext transfer protocol" (HTTP) is the language for the transfer and display of the information. Taken together, this trio takes your web browser software to the specific data records you have requested, pulls the information up, and then returns and displays that information for you.

The internet made things possible. The web made it usable.

Rewriting the Rules

Moving network routing activity away from a central point and distributing it at multiple points closer to the network's edge changed the nature of networks and ended the century-long run of Theodore Vail's vision.

An analog telephone call required a continuously connected circuit running through a switching center that kept the lines open for the duration of the call. A TCP/IP transmission breaks whatever is being transported into packets of data that travel to their destination by whatever network routes are available, and then reassembles those packets at the other end. The result: IP dispenses with keeping a circuit open, while replacing the switch at the center of a starburst pattern with a cobweb of connected routers moving tightly bunched packets at lightning-like speeds.

In other words, the move from requiring a dedicated circuit to a shared architecture that utilizes every millisecond of capacity destroyed the foundation of network inefficiency on which Theodore Vail had built AT&T.

Because digital transmission seeks out microseconds of unused capacity on multiple paths and fills it with packets of data pressed cheek-to-jowl to other unrelated packets, the cost to carry each incremental piece of data is virtually zero. Vail's economics of inefficiency no longer hold.

The relative absence of transport cost is illustrated by how Voice over Internet Protocol (VoIP) services such as Skype can offer "free" phone calls from computer to computer regardless of distance covered. A traditional phone line would require unused capacity waiting for someone to make a call and then dedicating an entire circuit to that single call. The longer the distance over which that circuit must be maintained, the costlier the call. When a telephone call is digitized, however, the packetized information is sent across the distributed fishnet network in an almost infinite number of route permutations based on the availability of tiny units of capacity in disparate networks. Because of the efficiency of high network utilization, the incremental cost of digital transmission approaches zero regardless of the distance traveled. One monthly internet access subscription from the phone company, cable company, or independent provider pays the cost of maintaining the available capacity, thus making each individual use of the network "free."[52]

Early applications of digital efficiency began to surface outside the AT&T system. New networks arose to challenge the Bell System—often using Bell's own lines. Led by men like Bill McGowan, companies such as MCI built their own long-haul networks as well as employed leased capacity from AT&T. Because they used digital technology, the upstarts could underprice AT&T, even when using Bell lines. AT&T, channeling Theodore Vail, tried everything possible to get the government to stop or otherwise control the new upstarts. But the digital cat was out of the bag.

The other characteristic of an IP network is that all information looks alike. Previously, as each new means of communication emerged, it required its own unique network. Radio and television, for instance, required a network different from that used to distribute telephone calls.[53] The advent of the lingua franca of IP brought about an era of convergence of previously separate networks where voice, video, and data were all the same—a collection of zeros and ones riding a common network to be converted into their final format by software on the receiving computer. An IP telephone call is a series of zeros and ones indecipherable from an IP bank record or an IP television video.

Perhaps even more powerful than its trans-platform capability, IP opened up a world of new applications and opportunities. The existence of a common IP platform allows digital information to be reorganized, repurposed, and redirected to create new or improved products and services. These new products are *iterative* in that they allow for new applications to be built on old ones (for instance, Facebook began as a platform for friends to post messages, but its all-IP technology has allowed the introduction of video and messaging). IP is also *compounding* in its ability to create something new by combining pieces of previously incompatible information into a new product (for instance, digital medical records have opened up new fields of medical research by allowing the records to be searched and related by treatment and outcome). *Creativity* also reached new heights because of IP technology that makes everyone with internet access a publisher and videographer. Finally, IP is *measurable,* creating new data measurement points every time it is used.

The combination of low-cost computing power and ubiquitous digital distribution redefined the nature of networks and their applications. The computing power that once had been locked away in special rooms and tended by its own priesthood was available to everyone. The centralized network that drew economic and social activities to a common point was dispersed, and commerce and culture followed. The information age had begun.

Connections

The technology originally developed for the purpose of improving national security has, fifty years later, enabled a new generation of threats to the security of nations, the sanctity of corporations, and the privacy of individuals. As one wag observed, the internet is "a lab experiment that got loose" to infect everything it touches.[54]

New networks have always introduced new threats to the traffic they convey. These threats, in turn, have stimulated new safeguards. Monks typically copied their texts only in daylight lest an overturned candle ignite not just the book, but the entire library. Railroads introduced onboard security to protect both passengers and freight from train robbers. Because telegraph wires could be tapped, elaborate cipher systems were developed to encode messages.

Paul Baran developed packet switching as a response to the threat of nuclear war. His goal was to ensure the ability to respond to an attack. Now the distributed architecture he developed is being used to flow forces in the other direction to enable cyberattacks on information, individuals, and infrastructure. The technology created to secure the old network became the basis for a new network that, as networks have always done, has opened new threats that demand new solutions.

That network security challenge is made manifestly more difficult by the hallmark of the new network: its distributed architecture that is open to all. Theodore Vail was able to secure the telephone network through incessant centralization of access, switching, and innovation. The security challenges of the twentieth century, from nuclear to chemical and biological weapons, also tended to be centralized and, thus, open to control. The national security strategy of "containment"—which protected the world after World War II—is possible only when the threat is centrally containable.

But containment is the opposite of the distributed forces of the internet. The twenty-first-century challenge is to reorient how we think about network security and to replace centralized containment practices with a decentralized dispersal of responsibility for our individual, corporate, and national security. Hiding behind firewalls and other static responses is about as effective against a cyberattack as the Maginot Line was in stopping the blitzkrieg.

In a distributed network the responsibility for protecting the network and those who use it is—like the network itself—dispersed. Individuals have a greater responsibility to protect their data as well as prevent unauthorized access to their computers. Corporations have the responsibility to use the new connectivity to establish collective, but distributed, defenses that share threat and mitigation information. Government must be a partner and facilitator in this highly un-government-like distributed response.

The history of networks until this point has been as a centralizing force for both the private and public sectors. As the new networks reshape economic activity in the opposite direction, it is necessary to rethink how we embrace network solutions to the new security challenges of a decentralized and open network.

Seven

The Planet's Most Powerful and Pervasive Platform

The village of Siankaba lies along the Zambezi River in the African Republic of Zambia. Home to 180 people and countless chickens, the village has no running water and no electricity. Aside from its inhabitants' huts, the predominant architecture of Siankaba consists of chicken coops built high on stilts as protection from nighttime predators. After a day of foraging through the village, the chickens, as if guided by GPS, return to the proper coop and climb its ladder to safety.

I wandered through Siankaba as the women were preparing dinner over open wood fires. Some of the men were setting up the evening's news and entertainment by hooking up an old radio to a car battery.

Adjacent to one of the dinner campfires, nailed slightly askew to a tree limb, was a crudely painted sign proclaiming "Latest Fresh Eggs on Sale." It made sense that someone identified on the sign as "Mrs. DR" would be selling the product of the ubiquitous chickens. What seemed out of place in this village, however, was the lettering squeezed onto the bottom of the sign: "Cell 0979724518."

In remote rural Africa, in a village of huts without running water or electricity, the cell phone is changing the basic patterns of life. Siankaba's inhabitants are part of the 95 percent of the world's population now covered by a mobile phone signal.[1]

Villages such as Siankaba have no water or electricity because the construction of the necessary infrastructure is prohibitively expensive compared with the potential users' ability to pay. And because the same market dynamics apply to telephone wires, the village was cut off from the outside world as well—until the advent of the wireless phone network. Nature's airwaves provide a low-cost pathway that enables the new network to be sustained by pay-as-you-use fees. Thanks to the economics of airwave distribution and low-cost phones, places like Siankaba can no longer be described as isolated.

When a villager in Siankaba can receive orders for eggs from a purchaser miles away, or can call a doctor about an ailing child, or can reach out to a

Sign in Siankaba, Zambia.

distant family member, life in remote villages in Africa has been changed forever. When Mrs. DR can be connected to billions of other mobile phone users located anywhere on the planet, life on all parts of that planet will never be the same.

In 2002, the penetration of mobile phones worldwide overtook the penetration of wired phones.[2] The telephone had been around for 125 years, yet all the telephone networks in the world combined had not become as pervasive as the mobile technology that had its first commercial trial only twenty-four years earlier, in 1978. In the intervening years mobile connectivity has soared to the point that many individuals have more than one device, and the number of commercial wireless connections is greater than the population of the planet.[3]

Today the mobile phone of 2002 is a museum antique; that it made phone calls without a physical connection was a wonder in its time.[4] In the new network revolution, however, wireless delivery and the internet have merged. The computing engine that started with Babbage is now a powerful processor in pocket or purse, and the universal network envisioned by Vail has become as ubiquitous as the air. Together they have created the most powerful and pervasive platform in the history of the planet.

The Path to Ubiquity

In 1873, the Scottish physicist James Clerk Maxwell published his "Treatise on Electricity and Magnetism," in which he hypothesized "that if an electric current were to surge back and forth through a wire very rapidly, then some of the energy in this current would radiate from the wire into space as a so-called electromagnetic wave." He called the wire from which the wave emanated an "aerial" or "antenna." Sixteen years later a German physicist, Heinrich Hertz, proved Maxwell's theory by generating electromagnetic waves in his laboratory. Hertz's name was subsequently applied to the unit of measurement of those waves.

It was the young Italian Guglielmo Marconi, however, who captured the world's attention by harnessing electromagnetic waves to send telegraph signals. In 1901 Marconi achieved the impossible by sending a wireless telegraph signal from one side of the Atlantic to the other. Five years later the Canadian-born inventor Reginald Fessenden transmitted an audio signal to ships at sea.[5]

The ability to transmit sound—including the human voice—without wires was the ultimate threat to Theodore Vail's concept of universal service provided by AT&T. Shortly after Marconi and Fessenden, Wall Street began to worry about AT&T's future. Who needed wires if the ether could deliver a conversation? Vail responded in January 1915. The AT&T board of directors appropriated $250,000 to develop a radiotelephone.[6]

Only nine months later, on September 29, AT&T engineers moved Vail's voice from his desk telephone in New York via phone lines to an antenna in Arlington, Virginia, where it was cast into the air and received as far away as Honolulu.[7]

In a congratulatory telegram to his chief engineer Vail wrote, "Your work has indeed brought us one long step nearer our 'ideal'—a 'Universal Service.' "[8]

The following year, in his 1916 annual report to shareholders, Vail, comfortable in his dominance of the technology, reassured those who worried about wireless competition. "The true place of the wireless telephone, when further perfected," he wrote, "has been ascertained to be for uses supplementary to, and in cooperation with, the wire system, and not antagonistic to it or displacing it."[9]

It would take sixty years for the technology to be "further perfected." The result of that development would belie Vail's assertion that wireless networks were "not antagonistic" to the wired network.

The first phase of mobile communications began in 1921 when the Detroit police department took the initiative to put mobile radios in squad cars. "Calling all cars" was followed in 1929 by ship-to-shore radio, which connected an ocean liner passenger directly into the Bell System.[10] During World War II mobile radios for police and emergency vehicles evolved into portable radio telephones the troops nicknamed "walkie-talkies" and "handie-talkies."

Thirty years after Vail's vision of a supplementary mobile telephone service, AT&T began offering the capability. On June 17, 1946, AT&T's Southwestern Bell subsidiary launched its Mobile Telephone Service (MTS) when a St. Louis trucker was connected directly into the wired Bell network. Shortly thereafter the service was rolled out to twenty-five other cities.[11]

The problem with MTS was the limited amount of available airwaves. MTS was a "high tower–high power" technology in which a multi-hundred-foot antenna blasted out a signal as far as possible. The return path to the tower also required a powerful signal, necessitating an eighty-pound transmitter-receiver in the car or truck. The mobile unit sucked so much power that its use would cause the vehicle's headlights to dim. Only a handful of individual channels were available to carry the MTS calls, limiting the number of individuals who could use the service at any one time. In all of Manhattan, for instance, the network could serve only about a dozen users simultaneously.[12]

The allocation of the airwaves was controlled by the Federal Communications Commission (FCC), the same agency that regulated AT&T's wire monopoly. In 1947, AT&T petitioned the agency to make more of the airwaves (technically called the electromagnetic spectrum) available for mobile telephones. Two years later the FCC allocated a few more channels. In a break from the "natural monopoly" concept, however, the FCC allocated half of the new channels to non-Bell entrepreneurs. It would be another thirty years before the Bell System tasted real competition, but the door had cracked open.

Also in 1947, researchers at Bell Labs began investigating whether there might be a technological solution to the laws of physics that limited the number of signals available for mobile communications. Drawing on the work of Rae Young, Doug Ring wrote "Mobile Telephony: Wide Area Coverage" and advanced the innovative idea that if the spectrum could be geographically divided into a honeycomb of hexagons, each on a different channel and operating at low power, then the common block of spectrum used for the "high tower–high power" solution could be subdivided into smaller noninterfering cells capable of serving more subscribers. It is this concept that is at the heart of modern mobile networks. The work was filed away and never published.[13]

For the next twenty years, the concept of a cellular network languished at both AT&T and the FCC. Ring and Young had conceptualized a breakthrough, but neither existing technology nor the necessary innovative corporate or regulatory vision was available to develop that concept. The need for portable computing power to handle both signal sensing and handoff to the next cell would also have to await the development of the microprocessor. The FCC continued to focus its spectrum allocation efforts on big blocks of airwaves for broadcasters. All the while, AT&T continued to exploit its "natural monopoly."

In 1958, AT&T petitioned the FCC for additional spectrum to be used for a mobile phone service. The agency sat on the request for over a decade.

As the federal agency responsible for the efficient and innovative use of the public's airwaves sat on its hands, another federal agency stepped in with a wake-up call. In 1968 the Department of Transportation hired Bell Labs to develop ways to implement pay phones on the Metroliner, the new high-speed train between Washington and New York. Bell Labs recommended dividing the 225 miles between Union Station in Washington and Penn Station in New York into nine cells. Just as originally proposed by Ring and Young, each cell had its own unique frequency. When the edge of one cell was reached the speeding train would trigger a sensor on the track, send a signal to a computer in Philadelphia, and that computer would hand off the call to the next cell. The Metroliner pay-phone service—the first cellular system—became operational in January 1969.

The Department of Transportation's radiotelephone leadership helped awaken the FCC. In the summer of 1968, the agency dusted off AT&T's decade-old petition for more spectrum. It took another two years, but in 1970 the commission asked AT&T to demonstrate how a workable car phone system could be developed using the cellular concept.

Yet there was skepticism about the airwaves being used for something as pedantic as phone calls. FCC commissioner Robert E. Lee, representing the antagonism that had held the topic back for over a decade, warned of "frivolously using spectrum" to enable people to call each other.[14]

In December 1971, AT&T formally submitted to the FCC a proposal for Metroliner-like frequency reuse. It would be another six years before the FCC allowed the company to construct a demonstration trial of its proposal.

I remember being at the FCC one day in the mid-1970s and seeing a demonstration of the cellular handoff. It was such a foreign concept that AT&T had constructed a demonstration model. A toy slot car was modified to follow a path embedded in a model landscape on a large piece of plywood. It was like a toy train layout with miniature trees and homes, except tiny lights had been buried in the landscape in a hexagonal pattern. As the car moved from one cell to another, the active cell would light up until the car reached its edge at which time the next-door cell would light up and the previous cell would go dark. It was a simple demonstration of the revolutionary concept of spectrum reuse through small, low-powered cells.

The FCC's delay, in part, was an echo of how earlier network innovations had been fought by those with whom they would compete or whom they might displace. Just as tavern owners and haulage companies fought to thwart the railroad's expansion, those with vested interests in the "high tower–high power" approach to mobile telephony—existing licensees, as well as their equipment manufacturer, Motorola—used the administrative processes of the government and court appeals to slow down the progress of the new network that would compete with them.

Finally, in 1978 the Illinois Bell subsidiary of AT&T began operating a cellular system of ten cells covering 21,000 square miles around Chicago. It would be another five years before the Regional Bell Operating Company Ameritech would begin commercial cellular service.[15] During that interlude, cellular service was launched in Tokyo (1979), Mexico City (1981), and cities throughout Europe.

The most significant development to emerge from the FCC's lengthy consideration of the rules for cellular networks was the decision to ditch

Theodore Vail's "natural monopoly." When the FCC finally allocated spectrum for mobile service, it did so to two *competitive* providers in each market. One of those licensees would be the local phone company, the other an entrepreneurial competitor.

Theodore Vail must have somersaulted in his grave! The "natural monopoly" was no more. The introduction of competition into the mobile telephone market accelerated deployment, stimulated innovation, expanded service, drove down prices, and ultimately spilled over into Vail's wired network.

At the time, however, no one understood the Rubicon that had just been crossed. The operating assumption that wireless connectivity was just as Theodore Vail had forecast in 1916—only an ancillary service—was illustrated by a 1980 study AT&T commissioned from McKinsey & Company. In response to the telephone company's request to forecast the demand for cell phones, the consultants forecasted there would be 900,000 cellular subscribers in the year 2000.[16]

Redefining a World without Wires

In 1992, as president of the Cellular Telecommunications Industry Association (CTIA), I traveled to the small town of Thibodaux, Louisiana, to recognize the nation's symbolic 10-millionth cellular subscriber. The McKinsey forecast had fallen significantly short of the mark. With eight years remaining until the century mark, wireless phone penetration was already more than ten times greater than the McKinsey consultants had predicted for the end of the century. By the time the millennium finally rolled around, wireless subscriptions in the United States were approaching 100 million, 100 times greater than the consultants' forecast.[17]

The 10-millionth U.S. subscriber was symbolic of the change the new technology was bringing to the way people led their lives and conducted their business. The designated 10-millionth subscriber was a large-animal veterinarian. At the celebratory luncheon in Thibodaux, she told the story of tending to an injured animal out in a field when her new cell phone rang with an emergency call from a client whose cow was in labor with a breeched calf. She marveled at how previously she would have been unreachable. Now, however, from another farmer's field she gave instructions and then rushed off to tend to the animal in distress.

It all seems so common to us now that the veterinarian would be reachable in an emergency; wireless connectivity has so rapidly changed our lives that it is hard to recall the unconnected isolation that was so recently the accepted norm. By 2012, this freedom and convenience were causing consumers to desert Theodore Vail's wired network to such an extent that more than half of all American homes did not have or did not use the wired telephone network.[18]

In the developed world, among the already connected, the introduction of wireless networks has increased productivity and enhanced convenience. In the developing world, among the never connected, it has changed the patterns of life itself. In Bangladesh a call to 7-8-9 on a mobile phone connects to the Healthline medical call center, where a doctor diagnoses symptoms and prescribes treatment. Calling the doctor is a simple application of wireless access, but in a nation with only five health care workers per 10,000 people, that connectivity is lifesaving.

In seaside India, checking a mobile phone before venturing onto the ocean can warn fishermen of weather over the horizon that could swamp their boats. After they return from a successful catch, a call from the boat tells the fishermen which port is paying the highest price. The fishermen's profits have increased by 8 percent, while the improved coordination of supply and demand has reduced consumer prices by 4 percent.[19]

Like the network revolutions that preceded it, the mobile network has produced innovative and unanticipated applications. The fee for the Bangladesh Healthline medical consultation (about 21 cents for three minutes), for instance, is paid by deducting it from the phone's prepaid airtime balance.[20]

The prepayment to the mobile operator for minutes of airtime has created a new asset class for individuals who have never seen a bank and never had a savings account. Airtime remittances use mobile minutes as a pseudo-currency that can be transferred between phones and exchanged for goods.[21]

In Senegal, another unexpected derivative of mobile connectivity has produced another unexpected benefit: text messages have become an impetus for literacy. Historically, African cultures were based on an oral tradition without much day-to-day need for literacy. Only a few in each village were selected for the "unproductive" effort of learning to read and write while all the others did useful labor. Daily crop, weather, and health information via text messages, however, is creating a new demand for literacy, as well as the means to practice reading skills.[22]

Nobel Peace Prize recipient Mohamed Yunus has argued, "The quickest way to get rid of poverty is to provide everyone with a mobile phone."[23] A World Bank study found that every 10 percent increase in mobile phone penetration in a developing country increased GDP per person by 0.8 percent.[24]

From Voice to Data

Events in the South African township of Soweto helped awaken the world to the outrage of apartheid. As I stood in the gate of a schoolyard in one of Soweto's shanty towns, the local women were lined up across the dirt road at the communal water spigot awaiting their turn to fill empty cans. Behind me, inside the school compound, the children could access the internet using wirelessly connected laptops. The juxtaposition was striking.

The Soweto laptops were part of the One Laptop Per Child initiative, which made available specially constructed and ruggedized computers capable of creating their own wireless mesh network in which each laptop acts as its own send/receive router. Each computer links through the air with the others and ultimately to an internet access point.[25] The laptops exemplify how most of the world's young people, whether in developed or developing nations, will have their first internet experience via a wireless connection.

Worldwide, the data traffic on mobile networks is greater than that of voice traffic, and it continues to grow.[26] Like the analog wired network, the analog wireless network was pressed into the delivery of data by "tricking" the network to carry digital traffic disguised in analog form. It wasn't until fourth-generation (4G) wireless technology began rolling out early in the second decade of the twenty-first century that wireless networks completed the transition to fully digital IP systems with broadband speeds. And because it is a packet-switched—rather than circuit-switched—network, costs are significantly below that of earlier generations of dedicated circuit wireless infrastructure, allowing its capabilities to extend even further.

The wireless revolution has not stopped with 4G, however. One of the most significant decisions of my tenure at the FCC was the identification of pieces of the airwaves to be used by fifth-generation (5G) wireless technology. While earlier generations of wireless technology evolved from voice to data (just as the wired network had), 5G was the first technology to be built from the ground up for the purpose of microcomputers talking to microcomputers. The 5G network will operate at speeds up to 100 times faster than 4G, with the capacity to handle a million transmitting chips per square kilometer, and with latency (the time required to deliver the data) as low as one one-thousandth of a second.

The new network revolution has come to pass: the convergence of portable digital processing with wirelessly delivered, IP-based services. It is the marriage of forces that began centuries ago. The computing technology that evolved from Charles Babbage's work to the microprocessor, the binary impulses of the telegraph, and Gutenberg's idea of breaking information into its smallest usable parts have all combined to redefine how we connect, and how we live.

Shaping Our Now

Chapter 8 explores some of the consequences of the digital wireless revolution. At this point, however, let's consider three real-time results and how they echo previous network changes: their impact on social patterns, their unanticipated results, and how change sparks cycles of concern and opposition.

The last network revolution of the railroad and the telegraph/telephone redefined the nature of individual interaction that had existed since the dawn of the agrarian age. Prior to the railroad, most people lived to produce for themselves or a small local economy. Even the principal cities were a shadow of what they would become as a result of rail connectivity. The notion of "community" was geographically limited to the family with occasional extensions to a neighboring market center.

The centralizing force of the railroad pulled people from a rural life of individual self-sufficiency to deposit them into melting pots of masses of workers and their families. With people piled into urban tenements, life spilled out onto stoops and streets to define an expanded sense of community built around not only physical interactions at the local market and at work, but also the fact that they simply couldn't get out of each other's way.

As wireless networks have untethered individuals, they have redefined the nature of community. The last network revolution produced a collective community. The new network is delivering what appears to be a contradiction: an individualized community.

It used to be that physical place defined everything from friendships to ideologies and bowling leagues. Wireless connectivity has made physical proximity irrelevant. To a wireless user, the relevant community comprises the people in their mobile device's contact list with whom it's possible to instantaneously connect, regardless of physical location.

In the process, the concept of being "alone" has evolved. Prior to mobile devices, being alone meant physical isolation. Now, however, "alone" has become "not connected." When a person's solitary walk through the park is interrupted by a ringing phone, that person is no longer alone. A teenager by herself in her room texting her friends is not isolated.

When Thoreau, complaining about the telegraph, said that "Maine and Texas, it may be, have nothing to communicate," he could not have been more off target.[27] New networks accelerate the desire to communicate.

The reality of the twenty-first century is that the *ability* to communicate makes it *necessary* to communicate.

Today the wireless network goes one step further to redefine what it means to communicate. First, the mobile phone removed the need to be in a specific location to exchange information: situating oneself at the front stoop, or the telegraph office, or where the phone wire came through the wall is no longer necessary. As a result, in the mobile world, "place" no longer exists.[28] Mobile technology makes it possible to be present without being in attendance.

The mobile device then adds another "nonplace" dimension by allowing us to be physically present with one group while involved with another. Interrupting a face-to-face conversation to answer a mobile phone call departs from the physical in favor of virtual proximity. Sitting in a meeting while checking email on a mobile device makes the user only partially present and erodes the importance of place even further. In the ultimate withdrawal from the physical, the mobile phone allows its user to consciously seek isolation from those with whom they are physically present. Mobile messaging usage is high in Japan, for instance, because subway riders, packed like sardines, retreat to their virtual communities, isolating themselves from the crowd around them by transporting themselves to connect to another, more intimate group.

The communities in which wireless users live are significantly smaller and less random than their nineteenth- and twentieth-century predecessors. One study discovered that about half of a person's mobile phone calls and text messages are sent to a limited group of only three or four people.[29] Such small communities, limited to the family and a small group of friends, echo the pre-industrial era.

Mobile connectivity can be Janus-like. On one face, small virtual communities of families and friends become emotionally closer because of constant connectivity. The other face reflects the ease with which an individual can retreat from the physical world into a wireless zone of comfort that reduces the creation of proximity-triggered relationships.

Beyond the changing nature of community, the history we have reviewed establishes that the transformational impact of a new network is not the primary network itself, but the unanticipated ways in which people put the networks to use. Printing could be logically expected to spread information, but its impact on navigation and commerce was less obvious. The railroad's unanticipated effect on urbanization became obvious only in retrospect. The telegraph obviously sent messages at near light speed, but its effect on the national spread of news or finance was unforeseen.

The historical pattern of nonobvious network effects continues in the new digital mobile environment. We have already seen how in developing parts of the world the mobile network has produced unanticipated consequences from pseudo-currency to the increased demand for literacy. In the developed (read: "already connected") world, unanticipated applications continue to redefine the mobile device's impact on our lives.

Consider, for instance, ubiquitous text messages from one mobile device to another. Today users send billions of text messages daily, yet it was never intended to be this way.

Included in the GSM standard for mobile devices was the ability to use the mobile network's control channel (the pathway that controls the call but doesn't carry the call itself) to send short alphanumeric messages. It was envisioned principally as a means for one-way communication from the company to the subscriber (such as "Your bill is due"). That changed when the functionality was discovered by Norwegian teenagers in the late 1980s.[30]

It seemed illogical that consumers would send a text message using the Short Message Service (SMS). First, the original form of text messaging required a seemingly endless number of keystrokes per word. The letter *c*, for instance, required hitting the phone's "2" key three times to toggle through *a* and then *b* before reaching *c*. The simple message "Running five minutes late," for instance, required forty-nine keystrokes on the mobile keypad. It was so illogical that the mobile operators never established a tariff to charge for the capability. And, on top of that awkwardness, the message was limited to 160 characters.

But "free" and "something my elders don't understand" were magic for teenagers. Not only could they "talk" as much as they wanted and pay nothing (at a time when voice calls were charged on a per-minute basis, often at rates of 25 cents per minute or higher), their parents didn't understand the technology, let alone the shorthand code developed to keep messages short (such as LOL for "laughing out loud"). Spread by the teens' powerful network, text messaging took off—and not just in Norway. By the time mobile operators awakened (and had to add network capacity to carry the influx), the volume of text messages numbered in the billions. As mobile operators began to charge for the service, it became an unanticipated, high-margin, big-profit activity.

Then the unanticipated struck again.

The original SMS standard imposed a quasi-digital peer-to-peer protocol on the analog mobile network. As mobile networks went digital, however, they challenged the primacy of the proprietary SMS capability. Competitive services using the open IP platform, rather than the closed SMS platform, began springing up—and, once again, those services were "free." Much as Skype rides the internet to provide "free" voice calls, independent texting services use IP-based messaging to bypass the mobile operator's SMS platform. The resulting relief to the consumer's pocketbook hit mobile operators' bottom lines hard. In the new network world, the unanticipated giveth as well as taketh away.

A similarly unanticipated impact of mobile has sent banks and credit card companies scrambling. As we have seen, airtime remittances denominated in minutes rather than currencies have sprung up around the world. In Kenya, however, mobile operator Safaricom hit pay dirt with its M-PESA payment program ("pesa" is the Swahili word for "cash," "M" stands for "mobile") that exchanged legal tender, not airtime minutes, between mobile phones. The customer registers with Safaricom for an M-PESA account into which cash is deposited. Within M-PESA the cash becomes an electronic replica of real money that the system then delivers like money transfers in the developed world. In 2012, one-fifth of the Kenyan GDP reportedly changed hands through M-PESA.[31]

Another unanticipated effect of the new network has been the reigniting of an entrepreneurial economy and at the same time a blurring of the line between home life and work life. By the beginning of the twentieth century, the railroad and telegraph/telephone had transformed an agrarian economy to an industrial economy in which industrial workers outnumbered farm workers. By the dawn of the twenty-first century, however, almost 60 percent of American males no longer worked on the shop floor but were engaged in information-based business.[32]

The network-enabled Industrial Revolution destroyed an economy of artisan entrepreneurs operating out of or near the family homestead. The village smithy, for instance, could not survive the competition from the distant factory's economies of scope and scale production. Our network revolution reverses that history to restore the role of the productive and creative individual.

In the economy based on hard-goods production, masses of workers were brought together in one place to mass-produce products. In an economy of knowledge workers, the network transports information to the person using it, regardless of that person's location. The result empowers individuals to make their contribution on their own terms.

As the new network enhances those individuals' connectivity, it increases their physical and economic independence. Claims adjusters, appraisers, sales agents, customer support personnel, auditors and bookkeepers, even R&D geniuses, are just a cross-section of knowledge-based workers who can now offer their services while operating wherever they want, independent of corporate structure. And not just services are enabled; artisan-produced goods, or kitchen-table companies, can reach worldwide markets right through their mobile device.

One unanticipated result of such economic restructuring has been the new network's effect on the family. When earlier networks moved workers into a centralized workplace they upended a tradition dating back to prehistory in which the family and the workspace were collocated. Our preindustrial blacksmith's forge was normally located in proximity to his home, and his professional and personal lives were intertwined. The factory changed all of that. Once people had been pulled together in urban ghettos, the breadwinner was taken to the factory floor for most of the waking day. In 1800, only 5 percent of men in New York City worked outside the home. By 1840, that number had swelled to 70 percent.[33]

The ability to be constantly connected to work is reversing the leave-home-to-go-to-work pattern. Work that can be done from anywhere, especially from home, is creating a "historical reintegration" that is returning society to the preindustrial era when work and the family were not physically separated.[34]

New Networks, Old Reactions

As history repeats itself with unanticipated results from new networks, it reiterates the resistance network-imposed changes produce. Whenever old institutions and the security of their conventions are threatened, a rearguard assembles to contest the innovations.

While working with the UN Foundation's efforts to use mobile technology to improve health care in developing countries, I witnessed the twenty-first-century reiteration of the Establishment's opposition to network-driven change. As we have seen, the ability to connect remote areas without a health-care delivery system with doctors located far away can save lives. But in one high-need African nation, the health-care professional association's response to such innovation was to add a new staff position—not for the purpose of facilitating such improvements but for challenging them. In a nation with a chronic shortage of health-care delivery, the position of "director of unnecessary technology" stood as a bulwark against the innovations made possible by the new network. It was the twenty-first-century reiteration of the nineteenth-century tavern owners and haulage companies who opposed the expansion of the railroad.

A similar narrative has emerged regarding the network's new infrastructure. In the nineteenth century, landowners fought to block the construction of rail lines across their fields. Today, that sentiment has meant opposition to the antennas that provide wireless connectivity. It has even become associated with an initialism, NIMBY—"Not In My Backyard." During the railroad construction period, opposition tactics included hiring bullies to attack engineering crews. The opposition today is, fortunately, more civilized, but no less intense.

The response of businesses confronted by the challenge of new networks also follows the old pattern. When early rail lines followed the same natural pathway through the Mohawk Valley used by the Erie Canal, the canal interests fought to limit the competition. Canal owners as well as barge operators, tavern keepers, haulage companies, turnpikes, and others affected by the expanding rails erected obstacles to protect their livelihoods. There was a similar response over a hundred years later when wired telephone companies tried to hobble wireless competition.

While the Bell companies had mobile assets they wanted to grow, they preferred that such growth follow Theodore Vail's 1916 claim and remain "supplementary to, and in cooperation with, the wire system, and not antagonistic to it or displacing it."[35] The competitive licensees created by the FCC, however, had the opposite idea. As CEO of the cellular industry's trade association during the industry's growth years, I was constantly refereeing the conflicting interests of the entrepreneurial wireless competitors and the incumbent wired phone companies.

One struggle echoed the earlier days of the wire-line business, when AT&T used its control over connecting to the network to constrain independent telephone companies. In the modern case, federal law required such interconnection, but the regulation didn't say at what price. The wired companies, for whom the interconnection fee from their mobile operations simply moved money from one corporate pocket to another, favored high fees that helped keep wireless prices high. The competitors fought to reduce the fees to a level that resembled the de minimus actual cost of connecting.

It was a war within the industry. When the insurgents finally prevailed over the incumbents, wireless could compete effectively with wired. The result was the wireless penetration we take for granted today.

History also repeats itself as new technology displaces the familiar and comfortable. Human nature dictates that deviation from old patterns creates a fertile ground for new suppositions; an intellectual worm of doubt gnaws at the public mind. In the railroad era, it was feared that the commotion of the speeding railroad would cause cows to stop grazing, hens to cease laying, and horses to abort their foals, while the locomotive's exhaust would cause birds to fall from the air.[36]

Shortly after the first telephone network was installed in Montreal, there was an outbreak of smallpox. The rumor took hold that the disease was carried by the new phone wires. Armed troops were needed to put down a violent mob intent on destroying the new telephone exchange. In January 1993, I was thrust into the middle of a similar panic when wireless technology was blamed for brain cancer.

David Reynard, a Florida widower, appeared on CNN's *Larry King Live* to discuss how he had filed a lawsuit alleging that his wife, Susan, had developed a fatal brain tumor as the result of the radio signals from her mobile phone.[37] The day after the CNN show, the stocks of mobile carriers and phone manufacturers plummeted, some subscribers canceled their service, and fear spread as the worm of doubt was released. More than twenty years later the worm of doubt still burrows away, though the U.S. Food and Drug Administration (FDA), the agency responsible for radiological safety, has reported that science has "not found sufficient evidence that there are adverse health effects in humans caused by exposure at or under the current radio-frequency exposure limits."[38] Multiple lawsuits, including the original Reynard case, have been rejected by the courts. Yet the worm of doubt still gnaws at the public conscience.

We have progressed from the days when new technology could only be explained as a tool of the Devil. Given human nature, however, change will always be difficult. The Baltimore clergy to whom Morse's telegraph was black magic and the complaint from the Ohio school board that the railroad was "a device of Satan to lead immortal souls to hell" still echo when a new technology confronts old conceptions.

Forward Connections

"The most profound technologies are those that disappear," Mark Weiser, chief scientist at Xerox's Palo Alto Research Center, opined in 1991.[39] As we look forward, the convergence of high-capacity wireless distribution with the inexorable growth in computing power outlined in Moore's law will make the network and its devices seem to disappear. The network that was built so that people could talk to each other has become the pathway for tiny computers to communicate without human involvement.

We tend to think of Moore's law in the context of how computing power has doubled about every two years. Now think of its other context: the reduction in computing cost. The manufacturing improvements that drove Moore's law compressed the cost of computing power, not just the size of the chip's circuits. It is this cost reduction that has enabled the explosion of microchips into everything around us. And that chip-enabled computing power will be connected without wires.

The new wireless networks—using both commercially licensed as well as unlicensed spectrum—will connect the hundreds of billions of chips that will all be awaiting commands or sending back information. The new network itself is microprocessor chips communicating with other chips.

The path of powerful wirelessly interconnected computing marches toward what some have labeled "digital dust"—an almost-infinite number of connected computing devices that form an "internet of things." Its potential is even greater than the original wireless revolution that connected people to speak to each other. The network that allowed humans to restructure their lives is now connecting a multitude of inanimate, interconnected microprocessors that will oversee new patterns of life.

The marriage of digital wireless connectivity with distributed processing power redefines the nature of common objects. Automobiles wirelessly report the performance of various systems and instantly notify emergency services in the event of a crash.[40] Houseplants send a wireless text message to "water me."[41] Our relationships with our own bodies are altered as wirelessly connected Band-Aid-like monitors report vital signs, while our pills call in to report whether they have been properly ingested.[42] Chips embedded in carpets even report what is going on above them.[43]

Hundreds of billions of interconnected "things" also create an unprecedented amount of digital information. In the next chapter we discuss how this activity is creating a new capital asset. The massive impact of billions of people being able to speak to each other will be overwhelmed by a transformational tsunami of inanimate "things" capable of exchanging information to influence how humans live their lives.

The new network revolution, driven by the most powerful and pervasive platform in the history of our planet, continues to develop. We are living amid a network revolution that promises to be of unprecedented size, scope, scale, and speed. It is neither the first time the networks underpinning our existence have changed nor the first time the results of those changes have made us question just what was going on.

Part IV

Our Turn

Look around, look around at how
Lucky we are to be alive right now!

—*Lin-Manuel Miranda,* Hamilton, *Eliza Schuyler verse*

"New networks challenge us to respond," I wrote in the prologue. The just-completed journey through the history of networks delivers us to today. Now it is our turn.

We have seen how new technology evolves in a Darwinian dance of innovation building upon innovation.

We have observed that while technology is the catalyst, it is the secondary effects of the primary network change that are transformative.

We have witnessed the struggles that occur when the immature effects of technology replace mature structures with a work-in-progress reset of the status quo.

We understand the narrative. We know the history. The only thing different is that now it is our turn to make history.

Eight

The History We Are Making

The headquarters of Facebook is a visual metaphor of the world we inhabit. The first thing that strikes you on walking in is that everything appears unfinished. The wall treatments and furniture are raw wood; even the giant iron beams that support the structure protrude in the middle of the open space, unpainted and unhidden, still bearing the spray-painted markings from their production and placement.

When I asked CEO Mark Zuckerberg about the décor, he explained it had a purpose: to remind everyone that their efforts are never completed. The unfinished surroundings reinforce to employees that they are in the midst of a revolution in progress.

Having concluded a review of the technologies that brought us to today, let's turn to a brief consideration of some of our own works in progress. What follows is an unscientific selection of some of the changes created by our new networks. As we reflect on these, we would do well to recall the experiences of those who earlier stood astride network-driven change. Their need to respond to the new networks—and the lack of obvious solutions—was no different from ours.

During my tenure at the FCC, I felt a bond with these histories. To remind me of this connection, I hung in my office a copy of the poster from the 1839 Philadelphia railroad dispute that we saw in chapter 3. Again, and again, I heard the issues raised by that poster replayed in twenty-first-century terms.

We share a common bond with the 1839 Philadelphians who were confronted by the social, economic, and emotional issues raised by the new network. If we had been around then, how would we have balanced the desire to protect successful social and economic structures and institutions while responding to the potential—only beginning to be realized—of the new railroad network? It is a history that echoes in the decisions we make today.

"History never looks like history when you are living through it," John Gardner observed.[1] Nothing is obvious. Our history will unfold according to how we deal with our new networks and their effects.

New in New Ways

What sets our time apart is how the networks that are triggering economic and social change are new in new ways.

That newness is rooted in the architecture of the networks themselves. The new networks' distributed architecture reverses the centralizing force earlier networks imposed on economic activity. At the same time, the decentralized applications of the distributed networks have recentralized economic power around the aggregated application of user information created by the network.

Previously, whether by boxcars or phone calls, networks transported everything to a central point to be transferred onward. We saw how Chicago became the Second City as the product of the plains was switched to rail lines heading east. Taking advantage of the transfer, Chicagoans milled the grain and slaughtered the animals before shipping them onward. Around such activities congregated support businesses that sold their wares to the millers, butchers, and railroad men.

Internet Protocol–based networks, however, do their routing not from a central point, but at multiple dispersed points, closer to the network's edge. Economic activity has followed that migration. From entrepreneurs who eschew the centralized lab for a garage, to the backroom artisan who can now reach a worldwide interconnected market, economic independence that was once crushed by network-driven centralization is being reborn thanks to new networks that push activity outward.

In many cases, the new network hub has been pushed to the ultimate edge: the individual. Mobile devices have assumed the role of a network junction; data comes in, is acted on, and then is shipped back out. But rather than the institutionalized traffic of previous networks, this in-and-out activity is highly individualized.

Our new networks also operate at near-zero marginal cost. In the old industrial economy, every time Ford built an additional car, the business faced steep marginal costs to acquire the metal, tires, labor, and other components. Today, however, every time Microsoft sells an additional copy of Word, it executes the virtually costless reproduction of an already assembled collection of zeros and ones, delivered at negligible cost over the internet.

When the old AT&T had to set aside a whole new end-to-end circuit just to handle one additional call, its marginal cost was high. But when the current AT&T digital network is chockablock with packets jammed cheek-to-jowl to achieve the greatest efficiency, incremental costs almost disappear. By one estimate, the online delivery of a one-megabit file (1 million bits) costs one-tenth of one cent.[2]

In the new network economics, the service provider must build the infrastructure (much as Ford must build an assembly plant), for which it deserves a fair return. But once the infrastructure costs are sunk, the marginal economics are quite different from those of the industrial era. It is the combination of zero marginal cost with open networks that has driven—and will continue to drive—innovation and growth in the twenty-first century.[3]

Perversely, the distributed architecture of the network has allowed some to recentralize economic activity. History's networks were centralized businesses that drove the centralization of the activities that used them. By distributing network functions—moving them "to the edge," in network parlance—the new networks have distributed to individual users the control over what goes in and out of their personal network connections while at the same time funneling into centralized databases vast amounts of information about those users.

The flood of information made possible by a distributed network precipitated an unforeseen phenomenon: non-network-based centralization. The driving force behind this new recentralization is the network's creation of a new capital asset: digital information. Previous networks *carried* assets to be put to work. Today's networks *create* new assets by the very act of carriage. Facebook, for instance, rides on the distributed physical network to aggregate information about its users and then to perform an algorithm-based analysis of that information so as to target those users for advertisements and information.

History's networks were "dumb": their job was to haul the asset to a point where its value could be realized. Today's networks comprise smart communicating computers that, by the very act of hauling information, create new information about usage and users. Every online activity leaves a digital footprint that is itself a valuable new piece of information to be bought and sold. It's as if the Post Office kept a record of every piece of mail you sent or received, and then sold that information.

The capital asset of the nineteenth and twentieth centuries was industrial production *facilitated* by networks. The capital asset of the twenty-first century is information *created* by networks.

Our new networks are also new in their velocity. The speed of networks inherently drives the speed of change. The time between the age of steam and sparks and the age of zeros and ones was half as long as the time between the printing press and the railroad, thanks to the accelerated speed of the network that was driving change. A locomotive could deliver the benefits of Gutenberg's product about ten times faster than when it traveled on horseback. Then the telegraph boosted that by another factor of ten. Today, tenfold increases in network speeds are puny artifacts.

Over the last few decades, network throughput has been increasing exponentially. During my tenure at the FCC, we established the threshold that defined "broadband" connectivity as 25 megabits per second (mbps)—representing an astounding eight-million-fold increase over the speed of the telegraph.[4] Yet even that threshold was almost instantly outdated as new network construction at gigabit speeds (one billion bits per second) became common.

The "new new" underpinning our future is the replacement of networks functioning as a centralizing force with those that act as a decentralizing force. By pushing network activity outward, digital technology has changed the economics of networks, sped up connections (and thus the rate of change), and delivered a new collection of network-delivered challenges, including the emergence of the network-using, service-platform-based centralization of digital information.

The Capital Asset of the Twenty-First Century

The writers' room at the Hollywood studio was spare: a long table and chairs extended from a wall covered in colorful five-by-seven cards charting the episodes for a television series' new season. Sitting at the table, I discovered how network-enabled data had driven the decision to turn one of my favorite fictional characters into a TV series.

Author Michael Connelly has written more than twenty best-sellers featuring LAPD detective Harry Bosch. Hollywood was interested, but the show had been in the purgatory that studios call "rework" for more than a decade. That was until Connelly took the character to Amazon for its Prime Instant Video service. It was a quick sell.

The author told me how surprised he was that, after years languishing in analog Hollywood, the digital gurus at Amazon moved so quickly to bring his character to video. He discovered it was all a matter of information.

The digital tracks created by the online purchase of a Connelly book, combined with information from other online activities, indicated Bosch readers were highly qualified prospects for Amazon's annual-fee-based Amazon Prime. In order to watch Amazon's videos, you had to be a Prime customer. What made this significant for Amazon was that Prime customers, in part because they receive free shipping, spend three to four times more on Amazon purchases.[5]

Connelly told me Amazon executives decided to produce a cop show because it would help them "sell razor blades" and other consumer products.

Welcome to the digital syllogism! Information about your digital behavior has a high value because it identifies likely future behavior and allows the users of that data to target how to influence that behavior. Every time a consumer does something online, that activity becomes an asset that can be monetized.

And "online" has an increasingly broad definition. Like virtually everything in the connected world, for instance, the smart TV in your home collects information about you. When television went all-digital in 2009, a computer replaced the cathode-ray tube behind the screen. When that computer connects to the internet to retrieve a video program, it also transmits valuable information about you upstream. You thought you were being sold a television, but it turns out that television is selling you.

The job of a computer is to count, manipulate, and store digital information. And computers don't just exist in products like your TV, PC, or smart phone. The modern digital network is a seamless collection of communicating computers. While old networks were "dumb" connections, the daisy chain of computers in IP-based networks creates analytical opportunities every step of the way. Not knowing precisely how to describe this phenomenon, conventional wisdom settled on the illustration of the internet as a fluffy cloud. The phrase joined the lexicon: information is transported, stored, and manipulated "in the cloud."

In the physical world of the last network revolutions, the railroad and telegraph hauled information from one point to another.[6] In today's networks of integrated computer connections, the act of transporting information creates new information about the content being transmitted (for example, "Tom is looking for a hotel in Paris"), as well as the context of that information (for example, "Tom is in Columbus, Ohio, at the corner of Broad and High, and has previously searched for online French lessons").

The knowledge and ownership of this computer-collected content, and its context, are at the root of the new network-driven economy. In previous network revolutions, information was static. While activity was always creating information, that information was largely inaccessible. The absence of processing power to collect and analyze the data, coupled with the inability to transmit it so that it could be combined with other data, meant the information went unused. Today, anything that touches a network becomes accessible information. And when information is put in motion to interact with other information, it creates even more new economic value.

Not surprisingly, the amount of digital information is exploding. Eric Schmidt, former executive chairman of Google's parent company Alphabet Inc., once estimated that "from the dawn of civilization until 2003 humankind generated about five exabytes of data"—that's approximately 5 million trillion bytes of data. He analogized that to "all the words ever spoken by humans to date." Schmidt liked to point out in 2010, that same amount of data was being created every two days.[7]

Schmidt's calculations have subsequently been surpassed. According to a 2017 study by International Data Corporation (IDC), we are creating 44 exabytes of new data on a daily basis. That's the equivalent of 3 million Libraries of Congress being created daily![8]

The creation of data is the manufacturing activity of the twenty-first century.

In the industrial era, capital assets were mined from the ground, transported to be fabricated into products, and then shipped to market. The capital asset of the new networked economy is information mined by and from connected computers and fabricated into new information products that are instantaneously available anywhere. Called "Big Data," this information tells a story of location, network logistics, and behavior—whether of inanimate objects or of you and me.

Every day the two jet engines on a Boeing 787 Dreamliner generate one terabyte of data—approximately 1 trillion bytes of data.[9] Sensors connected to microchips constantly monitor and report everything the engine does. Some of the information is networked to the plane's computers to combine with other data, such as weather information, to recommend the optimal altitude for fuel efficiency. Other information is sent to a satellite to relay performance data to the ground, while different information is stored for later download and analysis.

For an airline, information that increases fuel efficiency and monitors performance is invaluable as it can mean billions of dollars in savings per year. Industrial companies such as GE—maker of jet engines, among other things—have embraced data usage to redefine industrial production beyond the nineteenth-century concept of building good products, to building the product and then subsequently managing its performance based on computing power incorporated into the product.[10]

Industrial activity is morphing beyond production to include high-margin services that utilize intelligent components assembled *into* the product to manage the product's output. Industrial system decisions that once were based on statistical probability can now be made with near certainty. Whereas conclusions were previously extrapolated from a sample, embedded chips now measure everything, and intelligent software analyzes the data. It thus becomes possible to measure reality rather than simply forecast it. Statistical analysis just might become the new Latin—something you must learn but seldom use—as connected computing replaces statistical probability with knowledge certainty.

The applications of connected computing extend far beyond industrial activities. Information that has always existed, but was never captured and analyzed, changes everything it touches. In health care, for instance, the electronic patient records kept by 80 percent of office-based doctors are now an unprecedented research tool. If data from a jet engine can show how to fly an aircraft, data from health records can similarly change how doctors treat patients.

By correlating the data across 1,400 patients' electronic medical records, for instance, researchers found that a heart drug called a beta blocker had the unanticipated effect of prolonging the lives of women with ovarian cancer. Similarly, data mining of health records demonstrated that a drug prescribed for heartburn increased the chance of heart attacks. Such lifesaving findings didn't require a big laboratory project; the science was sitting there, unobserved, until the conversion of doctors' notes into digital information allowed metadata analysis, which in turn led to new medical breakthroughs.[11]

For a consumer-facing company, access to such Big Data is the key to solving a long-running riddle. Retail pioneer John Wannamaker is reputed to have said that 50 percent of his advertising was useless, but he didn't know which 50 percent. The new networks have created hundreds of billions of dollars in corporate value by using metadata to answer just such questions.

Google knows which product to sell you by estimating information about you based on your queries. Facebook knows because you told them about you, and then they augment that knowledge by monitoring your online behavior. The network that took you to Google or Facebook can watch all the unencrypted data as it passes and enrich it with real-time information such as your location, and even whether you are standing still or in motion.

And more data points are on the way. Microchips in jet engines are one example of the tens of billions of microchips being deployed in the internet of things. They will collect data and report on everything from the load in your trash can, to your driving performance, to your pharmaceutical consumption. Knowing this information, the networks will then carry back commands to the sensors to help the activity perform more efficiently, including telling you what to do.

Once again, networks are driving a new economic model. This time, however, it is for a different kind of product. The network revolutions of the nineteenth century enabled industrial production at scale. The networks of the twenty-first century are enabling the use of information at scale.

But What about My Privacy?

But what happens to my privacy when the digitally networked world knows everything about me?

Mark Zuckerberg once observed that privacy is no longer a "social norm." "People have really gotten comfortable not only with sharing more information and different kinds, but more openly and with more people. That social norm is just something that has evolved over time."[12]

Privacy has always been a relative concept. As America expanded westward in the eighteenth and nineteenth centuries, homesteads had the privacy of isolation. Yet in small frontier towns everybody knew about everyone else. Privacy was a casualty as the Industrial Revolution drew people into jam-packed tenements with thin walls and open windows. Then, after World War II the movement to suburbia began a return to geographic privacy protection, while in the cities the commercialization of air conditioning allowed privacy behind shut windows.

That privacy is an evolving concept relative to technological innovation can be traced back to—of all things—the box camera. In the late nineteenth century, George Eastman began selling a portable camera. It created a furor about personal privacy because anyone could record anyone else's images and actions without the other person's knowledge or permission.

In a seminal article, "The Right to Privacy," written in response to the camera's intrusion, Samuel Warren and Louis Brandeis argued, "Recent inventions and business methods call attention to the next step which must be taken for the protection of the person, and for securing for the individual . . . the right 'to be left alone.' " They went on to warn that "numerous mechanical devices threaten to make good the prediction that 'what is whispered in the closet shall be proclaimed from the house-tops.' "[13]

Louis Brandeis subsequently became one of the great Supreme Court justices. His concerns seem quaint in an era in which smart phones generate billions of pictures daily. But replace his term "mechanical devices" with "digital devices" and "whispering in the closet" with the digital utterance in an email or website visit, and it delivers us to our own question about the "next step which must be taken for the protection of the person."

How we respond to the new networks' ability to know everything about us will be one of the great cultural challenges of the twenty-first century. This response is made more complicated by the "privacy paradox"—the difference between expressed concern about private information and actual behavior. One study even suggests that society has moved beyond the paradox, that a majority of Americans are resigned to the fact they have already lost control of their information.[14]

Factors as diverse as age and nationality affect privacy perceptions. Millennials appear more accepting of the idea that information is known about them than do baby boomers. In the European Union, privacy is specifically identified as a civil right guaranteed to all, while in the Bill of Rights and the U.S. Constitution, the word "privacy" never appears.[15]

Yet networks have always posed a privacy threat. The founding fathers feared their revolutionary correspondence was being read. Messages sent via telegraph during the Civil War were often sent in code in case the enemy gained access to the line. Telephone calls were first held on party lines available to nosy neighbors; subsequent private lines were vulnerable to wiretaps.

The new networks add a compounding dimension to privacy. Historically, network privacy has been based on protecting the content of a telegram or a phone call, as well as protecting the information necessary for the network to operate, such as routing instructions and account information. Laws and regulations protected both the content and the context of communications. The extra effort required to access this information also served to protect it.

But what happens when the degree of difficulty to access your private information disappears, when the network is a connected chain of computers with the capacity to analyze and store digital information, and that information is a capital asset essential to a large segment of the economy?

The business models of early internet companies did not permit waiting around for gradual growth through subscriptions. As a result free distribution supported by advertising became the primary source of revenue. In an effort to solve John Wannamaker's problem, the early internet companies introduced cookies and other user identifiers: from that point forward, the ability of computers to count, calculate, and store information, coupled with the networks' ability to deliver that information with increasing granularity, has meant that tracking individuals is the default business model of the internet.

If you are not paying for an online product, then you *are* the product as information about you is collected and monetized. Even when you are paying for something like online connectivity, you are being monitored and monetized.

The smart phone has deepened the development of a revealing portrait about each of us. That wonderfully helpful technology is an unseen witness that constantly reports what we are doing. Your phone knows the basic information about who you are, where you live, and what you are doing online, but it also knows a great deal more about each of us.

When a smart phone communicates with a home Wi-Fi network, it also can report on how many people are in a room and even where they are sitting. When you are on the move, your smart device knows where you are, whether you are stationary, walking, running, or driving—and from where and in which direction. Because of a built-in altimeter, it knows the floor of the building you are on. It knows your likes, dislikes, and preferences; the books and news you read, the music you enjoy, and the TV and movies you watch. And, in the ultimate intrusion, your mobile can be turned on remotely without your knowledge to allow someone to eavesdrop and listen for key words, much like some email services collect key words in what you write.

As we've discussed, one of the principal effects of the new network is to distribute activity to the ultimate edge of the network, the user. As these in-out decisions of a network hub move to the individual, they are personally trackable and identifiable. As editorial creation and commerce move outward to become an individualized network hub, they also remove the ability to hide in the crowd. While the industrial age may have eliminated the privacy of the agricultural homestead, urbanity nonetheless allowed a degree of anonymity. That ability to hide among the congregated masses, however, evaporates as the network moves activity outward and tracks it to the individually identifiable user.

And the information never dies. As it is virtually costless to store a bit of data, there is little reason *not* to store as much as possible. Similarly, it is very inexpensive to search for and summon a piece of data and relate it to other pieces to enrich an individual portrait that can then be sold. As a result, our past begins to affect our future. The storage of information about our past behavior has an impact on our present and future by establishing expectations about us. Commercially, this detailed portrait is amazing and quite possibly helpful. Used by others (friend or foe) who want to know about each of us, it can be terrifying.

When such information is centralized and controlled, whether by a network or by a service platform like Facebook or Google, it can be weaponized to control an otherwise free market. By creating a data bottleneck, those who control the information can disadvantage traditional local businesses such as newspapers and independent merchants that don't have such access. Since data is the key currency, those who possess it can use their dominance to crush new competition and leverage themselves into new unrelated businesses. And since artificial intelligence is really nothing more than the manipulation of databases to reach a highly probable conclusion, control of those databases means control of the future.

We are not prepared to deal with the information explosion around us. Laws and regulations were designed in different times; networks are regulated by one federal agency, the Federal Communications Commission, while those delivering services are overseen by another, the Federal Trade Commission, with different legal expectations. One of the most challenging situations I faced at the FCC was how to apply a law written in 1934 and updated in 1996 to protect personal privacy in a world where information is a capital asset.[16] The conclusion we applied to network providers was based on three cornerstones: The consumer should be told what information the network is collecting about them and how it is used. Knowing that, the consumer has the right to choose whether to allow the information to be used in that manner. Finally, whatever information was collected must be securely protected from unauthorized use.[17]

Privacy is the conundrum of the twenty-first century. Using our information makes things run better and offers each of us new personalized services—but at the cost of individual privacy and marketplace competition. We have seen privacy expectations evolve over time. We must now determine those expectations in a digital world.

Horizontalizing Work

Before the railroad destroyed artisans in favor of industrial mass production, two blacksmiths, working their way through eleven separately identifiable tasks, could produce a plow in approximately 118 man-hours. The fate of these craftsmen was sealed, however, by a factory floor that produced the same plow in just 3.75 man-hours by assigning fifty-two men to perform ninety-seven distinct tasks.[18]

Achieving such productivity gains meant transforming individuals into cogs in a corporate machine governed by rules and processes. Two blacksmiths working side by side were a self-managing proposition. Production at scale, however, necessitated a hierarchy of management supervision to oversee and track multitudes of workers and their output.

Management hierarchy was another legacy of the railroad. As the first big business, the railroads recruited managers from the only other large institution of the time, the U.S. Army. The result was centralized, command-and-control management (ever wonder how military terms such as "division" came to be applied to corporations?).

Centralized management further expanded when the telegraph allowed supervision of multiple locations from afar. The result was that what happened on the shop floor, or the newspaper's front page, or the researcher's lab bench, was overseen by someone "upstairs," often multiple "upstairs."

The new networks stand management hierarchy on its head. Where the production of hard capital assets in centralized locations enabled a stratified corporate structure, information-based activity works in the opposite direction. As work moves off the shop floor into the cloud, coordination via hierarchy is being replaced by coordination via clicks to links and apps. In the process, the nature of work is returning to its preindustrial artisan roots.

Historically, artisans created the middle class. Johannes Gutenberg, for instance, was born into the developing economic and social stratum between the nobility and serfs, thanks to his father's skill as a goldsmith. Acting as independent contractors outside the vertical hierarchy of feudalism, these individuals performed specialized value-adding work on leather, metals, and wood. The middle-class artisans of the new millennium follow the same horizontalizing value-adding model, only they work with information.

The distributed network that gave us the "connected cloud" has enabled a "human cloud" in which individual economic activity is distributed just like the network that connects it.

Some economists predict that by 2020, about half of all American workers will be, like the artisans of old, freelance or contract workers.[19] Research in 2015 found that 22 percent of adult Americans had offered at least one service in the sharing economy exemplified by Uber and Airbnb, and for one-third of those individuals, that work was their primary or major source of income.[20]

MIT professor Thomas Malone coined the term "e-lancer" to describe the decentralized, task-based employment that is replacing the centralized, production-based employment of the last network revolution.[21] Dr. Nicola Millard of British Telecom creatively tagged this the "Death of Dolly," a reference to Dolly Parton's hit song "9 to 5" and the corporate employment structure it no longer represents.[22]

My daughter and son-in-law work for Salesforce, a pioneer in cloud-based services. Theirs is hardly a nine-to-five job. The cloud-based integration philosophy Salesforce sells to its users is practiced with a vengeance internally. Neither of them goes to a corporate office because cloud functionality means they can work anywhere, and always-connected means always available, whether to a client or a colleague. It is that nonintermediated collaboration with clients and colleagues that is most hierarchy-destroying. The new networks have returned us to the kind of collaboration among artisans that was snuffed out by the imposition of industrial hierarchy.

Nonhierarchical activity is also waste-reducing activity. Working free of office overhead or long commutes reduces the demand on resources. And the sharing economy, by definition, is less wasteful. An automobile typically sits unused more than 90 percent of the time in order to provide spurts of convenience a ride-sharing service could otherwise provide.

Yet the "Death of Dolly" has also hollowed out traditional paths to economic opportunity. For the last several decades, economic activity has been directed toward making things cheaper. As a result, employers exported jobs to low-wage countries while activities that stayed at home replaced people with robots.

While the new networks have been complicit in the "make it cheaper" market, they are also the path to new opportunity. "Make it cheaper" is being replaced by "make it smarter."[23] The products of the future are interconnected "smart" products—from smart cars, smart buildings and homes, to smart manufacturing and smart farming.

The challenge of a "smart" economy is "smart" workers. The skill set of many who lost their jobs in the early twenty-first century reflected a mismatch with the needs—and opportunities—of the modern economy. But the creation of widespread skills to serve new activities is a solvable problem we have seen before in history. The Industrial Revolution demanded workers with skills beyond those necessary for agricultural production. Those skills were delivered by an education system that applied industrial processes in twelve annual educational steps. Today we face the opportunity to redefine the concept of education to incorporate digital-era processes into the preparation (and re-preparation) of workers for the digital economy.

The new network that created the information economy is also the pathway for a new educational paradigm, as we will explore in the next section. A key concept of the digital economy—an iterative process to constantly improve the product—is as applicable to human capital as it is to the software that runs this new reality.

In the software world, a product release is just a beginning, followed by constant updates and improvements downloaded as they are developed. This iterative, agile improvement can be applied to the challenges of today, including retraining those whose industrial jobs have been eliminated for "smart" careers.

I saw the power of the new network to aid the continual expansion of individual skills in Hazard, Kentucky, where laid-off coal miners were learning software coding at the local community college. When fiber-optic cables brought the internet to coal country, local leaders and entrepreneurs—whose twang turns the word "hollow" into "holler"—began promoting the idea of "Silicon Holler," where the drive and determination that once took a person to the coalface could be applied to software coding.

The railroad network created coal country by hauling the natural resource to factories. Now the high-speed broadband information network is turning Kentucky coal miners into code miners to build and export the resource of the twenty-first century.

Retraining for a twenty-first-century skill set is a heroic effort. But the constant changes in the digital world also mean that training is no longer a one-shot process. Iterative education, where learning never stops, promises to redefine the workforce. Randall Stephenson, CEO of AT&T, told his employees that those who don't spend five to ten hours a week learning online "will obsolete themselves with technology." He warned, "There is a need to retool yourselves, and you should not expect it to stop."[24]

When simply keeping pace requires five to ten hours a week of ongoing education, a new reality begins to challenge old assumptions, including even the forty-hour workweek. During the New Deal, five eight-hour days replaced the sixty-hour week of six ten-hour days. Increased productivity permitted the week to shrink to forty hours. The digital economy's need for continual training may lead to an even shorter workweek, or to a week that includes training as a part of work hours.

The tendency of our networks to create nonobvious applications has also opened the door to new employment opportunities. Web applications, for instance, make work possible for those whose circumstances inhibit traditional nine-to-five employment. Want to generate income by offering doggie day care at home? The Thumbtack app promotes your service. Have a driver's license and a few hours? Uber makes you a taxi driver. Have a disability that hinders your ability to get around? Multiple online sites open new opportunities for work at home.

Inherent in the expansion of noncorporate work is also a new challenge to the maintenance of a social safety net. In Gutenberg's time, independent artisans sought security by organizing into guilds. While the guild system provided a structure of standardized processes and accreditation for independent producers, it also provided a degree of security against adverse events that could cripple a decentralized, independent contractor.

Through member contributions to a common fund, the guild economy helped provide security to its members. Members' hardships were answered with a withdrawal from the fund.[25] The term "guild," in fact, derives from the Old English word for "gold—"gild"—in reference to the gold in the group's treasury.[26]

As industrial scope and scale activity choked out artisans, workers sought security by banding together into unions. We have seen how the first big business, the railroad, drove the creation of organized labor as a countervailing force. Subsequently, through law and labor agreements, corporations became the vehicles for the delivery of worker security programs such as pensions and health insurance.

As large corporations fade as the principal provider of employment, however, the security needs of newly enfranchised information artisans do not go away. In an economy of on-demand individual contractors, the worker security that law and labor negotiations created in the nineteenth and twentieth centuries are being challenged to evolve. The increase in noncorporate employment brings the need for a new dispensing agent for worker security.

The ability to mitigate risk by collective action that began with guilds and fully developed in the industrial era now seeks a new vehicle. While never the focus of the debate surrounding it, the Affordable Care Act (more widely known as Obamacare), which managed risk-sharing through insurance exchanges open to all, was an early manifestation of this principle.

New types of labor organizations such as Coworker.org and the Freelancers Union are emerging. Some traditional unions have created an "hour bank" where an independent worker overpays monthly benefit premiums in an effort to build up a "bank" of surplus payments that can be drawn upon to cover the cost of benefits when they are not working. The creation of exchanges like those Obamacare established for health insurance can serve a similar purpose for disability and workers' compensation insurance.

As new networks disrupt traditional assumptions about the structure of work, they have expanded individual opportunity in a manner that recalls the days of artisans. Individuality has replaced rules-based hierarchies and, with it, the security that accompanied such hierarchy. In their return to artisanal activity, workers in the digital economy face similar protection and security challenges as did workers in earlier network revolutions.

How We Learn

The students at James Robinson High School in Fairfax, Virginia, receive an education unlike anything I ever experienced. Observing a science class, I watched the progress on their year-end projects. No papier-mâché volcanoes here! Each student had a laptop connected to the internet by Wi-Fi. The three-student teams were collaborating on a PowerPoint presentation that would be their final report. All of them were doing individual research and then contributing to the report, which was housed in the cloud.

As each team worked, the teacher monitored their activity by logging into the cloud. There she could also leave notes critiquing their progress—comments each student would see, along with the other students' replies. When a student or group appeared to need more hands-on support, the teacher could identify the need in real time and go to the students directly.

Networked computing, a teacher at another school told me, had changed the nature of teaching from the "sage on the stage" reaching a heterogeneous class with a common message, to the "guide by the side" helping each student learn at his or her own pace.

The education system with which most of us are familiar was developed in the nineteenth century to feed qualified workers into the industrial beast. The goal was to instill skills necessary for the factory floor, such as the ability to understand an instruction manual or make calculations. Mass-production techniques that had worked in the factory were applied to the classroom. Raw material went in, was improved at various points with specific processes, and emerged a finished product. In the factory, the production of a plow was ninety-seven specific tasks; in school, the production of a qualified individual was twelve grades.

Applied at a central location where scope and scale economies could be brought to bear, the cost of educating a student declined to the point where free and compulsory schooling was universal in developed countries. Industrial-age education was a huge success, but the industrial activities the educational structure was designed to support no longer drive the economy. The networked economy introduces new sets of skill requirements. Fortunately, it also introduces new means to deliver necessary training.

The developments we have just discussed—the changing nature of work and the use of information—manifest themselves in the evolution of education in two ways: the *what* and the *how.*

What students are being taught today often follows the old industrial age model, just updated for the twenty-first-century version of the factory floor. Nineteenth-century education was designed to impart the basic reading and math skills necessary to operate on the factory floor. The current educational emphasis on STEM fields—science, technology, engineering, and math—similarly focuses on the skills necessary for the twenty-first-century "factory" even though the factory "production line" may require sitting in front of a computer.

We will shortly discuss the impact of computers replacing jobs in the digital economy. The jobs that are most susceptible to elimination are those in which rote activities can be replicated in hardware and software. The jobs that are less susceptible to automation are those that require the uniquely human skills of common-sense reasoning and understanding the human condition. Broadly defined, we call these human characteristics "creativity" and "the arts."

In the digital era, employment opportunity relies on doing things computers can't. Zeros and ones can't think creatively. The educational system must prepare us—throughout life—to exploit that advantage. Our only protection is an educational system oriented less toward teaching the skills of a second-rate robot and more toward instilling the skills of a first-rate human, one that computers can't replace.[27]

The *how* of the evolution of education begins to look like the new network itself. Just as work has moved from the centralized shop floor, in which the individual was a cog, to a distributed environment in which the individual can be a hub commanding the inflow and outflow of information, the new networks create a similar opportunity for flipping the educational process from mass to individualized.[28]

When students are individual information hubs, as opposed to a single unit buried in a larger group, they can command their lessons. Rather than sitting in a classroom trying to grab on to a concept as it flies by, the "individual hub" student can call up topic-specific online presentations. A teacher in front of a large classroom can't keep reviewing one point for the benefit of a single student, but when the student is a learning hub he or she can review an online explanation as many times as necessary. And if it still doesn't work, the student can find a different teacher, that is, a different online presentation on the same topic.

In 2006 Salman Khan, a young hedge-fund analyst, began tutoring his cousins in math over the internet. Ultimately, he put videos of his tutorials on YouTube, where they were so popular he quit his job and created Khan Academy, a digital learning company. His goal, he told me, was to allow anyone, anywhere, "to learn almost anything—for free." Today, tens of millions of people have taken him up on the offer by accessing thousands of online lessons. The lessons have their own drills, and through one-on-one interaction with the student, they allow the kind of focused instruction that is difficult in a large classroom.

Connected learning also produces the Holy Grail of the twenty-first century: data. It is estimated a student produces 10 million data points in a single day.[29] This information was always being produced, but schools and educators lacked the capability to either capture or process the data. How many times did the student repeat the presentation? Which points needed to be reviewed multiple times? What is the best time of day for learning a specific topic? What is the optimal amount of time to spend on new information? When Big Data comes to the classroom, it can help teachers teach as well as help administrators measure progress. Most important, it can help students learn.

Since the middle of the nineteenth century, education has forced the student to adapt to the system. The new networks make it possible for the system to adapt to the student.

Simply inserting the new network into an industrial age structure is not a solution, however. Networks change pedagogical processes as well as what needs to be learned. Industrial age schools focused on reading and math because of the need for workers to be able to read instructions and execute basic computations. While such skills remain necessary, the networked economy has redefined them.

In an eighth-grade English class at Edna Brewer Middle School in Oakland, California, I was initially surprised to see students working on what hardly seemed to be "English." Each student was incorporating teacher-specified modifications in font, color, style, and imported graphics to create a page on their computer screen. This was an English class? Wasn't this the kind of thing that should be covered in art class or the computer lab? Why was it in English class?

Then I realized two things that made these students' educational needs different from my experience. First, while they are growing up digital natives, students still need training in computer skills (just as growing up a native English speaker doesn't mean I still didn't need to learn English skills). The development of an intuitive relationship with networked computing comes through repeated interaction with networked computing—even in English class. Second, English class exists to help students learn to express themselves. For these students to be able to take advantage of how the network has become the greatest self-expression tool ever known, they will need to know how to create on the network, including these online formatting skills, and it will need to be second nature to them.

The ability of schools to access the new networks was one of the first issues we dealt with in my time at the FCC. As the agency responsible for supporting the connectivity of the nation's schools and libraries, we discovered that almost two-thirds of American schools and libraries were stuck with twentieth-century networks in an era of twenty-first-century needs. Schools serving 40 million students simply did not have high-speed digital fiber links to their buildings to carry the traffic necessary to allow multiple simultaneous internet users in every classroom. We also discovered that although Wi-Fi might be available in every coffee shop and McDonald's, even schools with high-speed fiber connections didn't have in most of their classrooms the Wi-Fi needed to put connectivity on the student's desk. And, of course, these problems were the most severe in schools serving rural and low-income areas. Our reforms created the opportunity for all students to have in-school access at their desks to the most powerful educational tool in history.

Previous pedagogy was designed to produce workforce skills necessary for routine, rules-based activities under hierarchical supervision. Today, our challenge is to produce a digital workforce for whom the network is an extension of themselves and an opportunity to apply abstract skills such as problem solving, analytical reasoning, and the communication of complex concepts. It is technology that makes the demands, and technology that enables new solutions. We just need to be smart enough to see and seize the opportunity for *all* students.

The Nature of Nations

"Networks are more powerful than nations," a senior U.S. State Department official once told me.[30]

The modern nation-state is the combination of geography and a sense of community. By perpetuating and distributing a common language, the printing press helped create that sense of national community. The early railroad and telegraph further knitted that community together for commercial activity and the information exchange.

Ultimately, however, the railroad and telegraph gnawed at the power of geography and expanded horizons across borders. Today's networks expand that orientation yet again, to a connectivity unfettered by maps and borders. When a single keystroke can circle the world in seconds, geography becomes irrelevant. In the process, the new networks have become a new challenge to the nature of the nation-state.

The modern nation-state is a seventeenth-century concoction. In 1648, after decades of war in pursuit of supranational empire, the European powers agreed to the Peace of Westphalia, in which they disavowed empire by affirming geographic national sovereignty. Exported to the rest of the world, national sovereignty defined the intervening centuries and was defended in battle.

By making geography irrelevant and creating new non-national communities, the internet has fragmented a world previously stabilized by national centralization. The result is a challenge to the nation-state from above by supranational networks and from below by super-empowered individuals and groups.

Significantly, the vehicle that challenges the nation-state was created outside of sovereign governmental structures. The internet is a set of technology standards developed by a non-national, multistakeholder technical community. Without sovereign direction, the networks of the twenty-first century were stitched together around common technical standards that were then imposed on the Westphalian world without permission.

The path to diminishing sovereignty began with the first electronic network. In 1865, the French government assembled representatives of other European nations operating state-owned telegraph networks. The attendees formed the International Telegraph Union (ITU) and agreed to a common set of regulations to bind their activities. It was the first supranational organization in which geography-based nation-states ceded sovereign authority to a common body.

The march from a telegraph union, to a coal union after World War II, and, ultimately, the modern European Union has been the pursuit of industrial-era scope and scale economics at the expense of geographic sovereignty. Such supranational structures formerly existed at the pleasure of the sovereigns. Yet in the new supranational digital network, technology is controlled by a community of technocrats who answer to no sovereign. This dichotomy creates awkward situations such as the European Union's effort to harness its member nation-states into a Digital Single Market (DSM) that nonetheless respects differing national policies, precisely as the distributed, lightning-fast network challenges the relevance of unique national policies.

Alongside such supranational realities, the new networks have empowered those wishing to challenge the state from below. Social media has altered the nature of the interaction between government and the governed.

In 2009, the sprouts of protest in Iran, dubbed the Green Revolution, were organized internally and then reported to the world by social media. Online activity was such an important tool that when the U.S. State Department discovered that Twitter—a major organizing tool for the protesters—would be briefly taken down for routine maintenance, it urged the company to postpone the work so as not to disrupt the protesters' ability to communicate.

As the Arab Spring of 2010–11 spread across North Africa, Cairo's Tahrir Square became the center of protests against the Egyptian government. Like the Green Revolution, protest organizing occurred over social media apps on mobile phones. Ultimately, the government shut down the mobile networks.

In the United States, the new networks have redirected political activity the same as they have economic activity—away from centralization—and with similar destabilizing results. Traditionally, centralized networks supported centralized media, which in turn supported centralized, hierarchical political activity. But decentralized networks empower individuals at the expense of institutions. From Barack Obama's groundbreaking use of social media in his 2008 campaign, to the ability of the Tea Party to set themes in Congress and campaigns, to the role of Twitter in Donald Trump's campaign and presidency, to the Russian government's use of social media to influence the 2016 presidential election, new networks have disrupted the role, functions, and importance of the political gatekeepers that previously provided stability.

The networks that expand citizen participation also undermine state sovereignty. The very networks that create economic opportunity, improve quality of life, and expand political participation also deliver the organization and management tools necessary for violent attacks on national stability.

If it seems, for instance, that twenty-first-century sectarian wars in the Middle East echo the religious wars that set Europe ablaze centuries ago, should we be surprised? Back then the printing press knocked down the walls that had constrained ideas and consolidated power. Today electronic networks allow insurgent movements to gain scale at speed and to combine military activity, social media, and terror attacks in a coordinated campaign. As Joshua Cooper Ramo has observed, "Network technologies do to terror attacks what gunpowder once did to projectiles. They make the impact larger."[31]

While in some instances new networks may threaten stability, in others they reformat the operation and the sustainability of government in positive ways. In Estonia, for instance, a citizen can conduct her entire relationship with the government online, including voting. There are only three governmental activities that cannot be executed online in Estonia: notarizing a document, getting married, and getting a death certificate (at the other end of life, a birth certificate is created digitally at the blessed event and sent to the family online).

Often called "the most wired country in the world," Estonia is the first "country in the cloud" where the state exists online.[32] While Estonia has the advantage of both its small size (approximately 1.3 million people, or 75 per square mile), as well as a greenfield restart after the end of the Soviet occupation in 1991, it is the ultimate manifestation of the new Westphalia: a state on the map whose functions are in the network cloud.

For other nations, the Estonia-like technological opportunity is clear, but the governmental legacy of previous network revolutions slows things down.

The centralized networks of the nineteenth and twentieth centuries created big business, big cities, big markets—and big government. As scope and scale became the hallmarks of economic activity, government provided an offset. To accomplish this, government imported into governing the same scope and scale structural concepts that ruled the economy. Hierarchical, rules-based oversight became the modus operandi of government just as it was for the operation of industry.

Yet, while industry can be ruthless in implementing new procedures in pursuit of profit, democratic government revolves around a much slower process that buffers change by providing all interested parties an opportunity to affect the outcome. Having run a bureaucratic agency, I understand that full and open participation is—and should be—of paramount importance. But it sure slows innovation and efficiency.

During my tenure at the FCC we worked to develop what we referred to as a new regulatory paradigm. In times of slower-developing technology, regulators could engage in top-down micromanagement of markets. Today's rapidly changing technology means that such an approach would not only slow innovation but would also be impossible to implement.

In this regard, we tried to learn from the disrupters themselves, the software developers. Early software development was linear and incremental; it was described as the "waterfall" approach because the development would slowly move downstream by stages until it was completed and went over the waterfall. Around the turn of the century, however, the waterfall was replaced by "agile" development in which the requirements and the solutions evolve in tandem. In agile software development the product is never done since it must always be responsive to new developments. Agile regulation means a similar articulation of essential principles and flexible enforcement if and as necessary, based on how technology and the market evolve.

Beyond the impact of networks on bureaucratic decision-making is their impact on the broader political process. The centralized economic and social realities that created the present structure of government are being replaced by new networks as surely as Amazon replaced the neighborhood bookstore.

When the founding fathers produced a republic rather than a pure democracy, they made the decision to create an intermediate layer of elected officials between the people and power. James Madison explained this in *Federalist* No. 10 as the "cure for the mischiefs of faction," caused by "some common impulse of passion."[33]

The centralized nature of the old information networks helped mitigate such passion by empowering curators such as editors or party leaders to structure debate by the way they assembled and organized information. Because the new decentralized networks are decidedly not curated, they replace such oversight with an unprecedented ability for individuals to organize and express themselves without permission or structure.

Created by the new networks, the aftermath of uncurated expression is a governing challenge for our era. "Winning is easy . . . governing is hard," George Washington tells Alexander Hamilton in the Broadway hit *Hamilton.* It is an observation relevant today as new networks expand the ability to organize and fight while making it harder to govern.

My friend Wael Ghonim, who used social media to launch and coordinate the revolt in Tahrir Square, defined the challenge. "The same tool that united us to topple dictators eventually tore us apart." Those who harnessed new networks in democratic expression in Egypt, Ghonim warned, failed to follow it with republican organization, in large part because organizing "anti" is easier than building "pro."

After bringing down the government, Ghonim says, "we failed to build consensus, and the political struggle led to tense polarization. . . . Our social media experiences are designed in a way that favors broadcasting over engagements, posts over discussions, shallow comments over deep conversations. . . . It's as if we agreed that we are here to talk at each other instead of talking with each other."[34]

In Egypt, such "talking at each other instead of talking with each other" allowed those who were organized, the Muslim Brotherhood, to hijack the revolution and replace one autocrat with another. In the United States, the result has been to gridlock collective republican (small r) decision-making.

We are not immune to the reality that networks shape the nature of nations, the operation of governments, and the role of the governed. Throughout history, networks have defined nations and enabled empires. Our new information-based networks challenge us to preserve hard-won republican ideals even as we embrace digital change.

Information Insurgency

On June 1, 1980, Ted Turner inaugurated Cable News Network (CNN). I was invited to speak at the launch. Struggling to articulate the groundbreaking all-news format in the context of what was then a world of video content scarcity, I described CNN as "a telepublishing event marking a watershed in information provision."[35] It was a bit over the top rhetorically, perhaps, but the point still stands: CNN started bringing the diversity of print publishing to video.

It all seems so curious today as 300 hours' worth of video is uploaded to YouTube every minute.[36] But CNN was the network-driven precursor to the untamed distribution of video information that we take for granted today.

CNN broke the news monopoly of the three major television networks by taking advantage of the expanded capacity of cable networks. In the process it introduced velocity to the news cycle. Previously, news waited its turn to access the limited capacity of broadcast networks. That meant an evening thirty-minute (twenty-two minutes without commercials) daily news roundup. Monumental events, of course, enjoyed "breaking news" status to interrupt regular programming, but that was rare.

Cable networks, unconstrained by the scarcity of broadcast spectrum, enabled CNN's all-news-all-the-time model. Events that would otherwise have waited went on immediately. As a result, the definition of what was "news" changed.

Immediately after the launch ceremony, Reese Schonfeld, the architect of CNN, took me into the control room to proudly point out live satellite coverage coming from a beach in South Florida in anticipation of a storm. Watching the sky and the tide to report on a storm that had not yet hit would previously not have been worthy of airtime; traditional television would have awaited the results. But with twenty-four hours of airtime to fill, the fact that the storm had not yet hit was news.

While CNN opened the time aperture that had previously constrained video reporting, it did not change the basic paradigm of curated news. Since the days of the first newspapers, the role of editor-curator had reigned supreme. Because of the limited space, someone had to make judgments about what was news and how much emphasis it would receive. For a piece of news to get on the front page of the newspaper, or on the TV networks' thirty-minute nightly recaps—and even on CNN's twenty-four-hour cycle—someone had to judge that it was more important than other pieces of news.

Such judgment perpetuated a hierarchy in which information flowed vertically from an authority. Whether that authority was the state, the church, or an editor, institutionalized information flowed downward from a creator-curator to the masses. Elaborate systems—called public relations—were developed to push information up to the curator in the hope it would be deemed worthy of being pushed back down.

As we have seen, technology invented the mass media. From Luther's "flying writings" (*Flugschriften*) coming off the printing press, to the Associated Press's harnessing of the telegraph, technology enabled innovative and unforeseen results. Now technology has disassembled the media as we've known it for the last century and a half, and reassembled it in a nonhierarchical, uncurated format, to take advantage of the total absence of any access bottleneck.

In a classic *Daily Show* segment, a faux reporter pointed out to a *New York Times* executive that what was in the paper had already been on the internet. "Why is aged news better than real news?" he asked.[37] Once again, the concept of news has been redefined. CNN may have meant coverage of an approaching storm was news, but today, when everyone is connected, it means everyone is a reporter commenting on and providing thoughts about the storm or any other topic. What is "news," therefore, has become an unsupervised personal judgment passed on at electronic speed.

According to the Pew Research Center, in 2015 the majority of Americans (63 percent) relied on personal postings on Twitter and Facebook as the source of their news.[38] This reality is not lost on the mainstream media. In the newsroom of the *Washington Post*, for instance, a huge video screen provides real-time reports of online activity. Those who produce the news are constantly reminded they don't work for a "newspaper" anymore; they are an electronic information service that receives and distributes information via a ubiquitous network.

The "shoe-leather reporting" that formerly was necessary to turn up stories, while still important, has been augmented (if not replaced in some circumstances) by ceaseless social media monitoring. Twitter feeds and Facebook postings are how news is created. And the big newsroom screen at the *Post* constantly reminds the journalists that social media both deliver them story leads and convey their writing outward to *Post* readers. When I was looking at the screen, Facebook and Twitter were the leading sources of readers coming to the *Post* (consistent with the Pew study). An important measure on the screen was also how long the online readers stayed with each story, and whether they went to another *Post* story after reading the one that brought them to the site in the first place.

While the *Washington Post* still performs an editorial function with the information it collects from social media, the unedited voice of the people that dominates the internet affects what is news without ever interacting with an editor.

One of the logical consequences of uncurated news from social media sources is an expansion of opinion. With everybody tweeting about his or her latest discoveries, the value of a scoop decreases. Thus, when product differentiation can no longer rely on scoops, it turns to opinion.

Journalism has always been opinionated. Early newspapers were political rags propagandizing for one side or another. "The golden age of America's founding was also the gutter age of American reporting," one commentator observed. Early newspapers were "conceived as weapons, not chronicles."[39]

It was the first electronic network, the telegraph, that lifted journalism beyond local bias and partisanship by introducing timely news from afar. This in turn accelerated the shift of news gathering from a tool serving the owner's political causes to a commercial activity. As a part of this commercialization, advertising revenues gained added importance. With the simple reality that more readers drove higher advertising rates, it became economically wise to offend as few as possible by offering balanced reporting. To maximize appeal to advertisers, therefore, the media attempted to practice objectivity, covering all sides of a topic and muting personal opinion so as to repel as few as possible.

My friend Ron Nessen tells a story about when he closed a 1966 NBC News report on the Vietnam War with an observation about the pope praying for peace. "I think we can all agree on the need for peace," he opined as he signed off. Upon emerging from the studio, Nessen was confronted by the president of NBC News. Even a brief hope for peace was out of line with the prevailing practices of journalism. "Nobody cares what YOU think, Ron," he was reprimanded. "Nobody cares what YOU think."[40]

It wasn't long thereafter, however, that cable TV "telepublishing" began to erode the economic rationale for objectivity by seeking profit in market segmentation fired by opinionated talking heads. "I think . . ." became the most often repeated words by the cable talking heads.

The movement of journalism to the internet picked up on and expanded that trend, but without the curatorial controls created by news gatekeepers. No longer costly to produce or difficult to distribute, text and video uploaded to the internet removed the hierarchical structure that previously policed media content.

The curator of traditional media was responsible for its veracity. Even when cable TV's real-time reporting led to errors, they were caught by editors. Live all the time meant cable news was "never wrong for long." However, when everyone has access to everyone else on the spur of the moment to produce a curationless "news selfie," our information flow begins to resemble the distribution of content before the centralization imposed by networks.

Tom Standage, who oversaw the digital transformation of *The Economist*, observed in his book *Writing on the Wall* that the last 150 years of centralized news control were an historical aberration. During the Roman Empire and earlier, news was a "social media" activity "in which information passe[d] horizontally from one person to another along social networks, rather than being delivered vertically from an impersonal central source."[41]

The telegraph helped institutionalize and commercialize local newspapers into a connected reporting apparatus based on rules and practices. Follow-on radio and television technology perpetuated the model of journalistic behavior controlled by gatekeepers who presided over scarce distribution capabilities. As the new networks dispense with such scarcity, the flow of information is, once again, being redefined.

This time, however, there has emerged a new kind of digital gatekeeper with a new economic incentive that is often in conflict with the desire for information veracity.

The financial success of a social media platform is determined by how long it can hold the user's attention in order to deliver advertisements. The average Facebook user, for instance, spends fifty minutes a day on the site.[42] To accomplish a long hold on users' attention, the platforms accumulate information about each user and feed it to software algorithms, the software recipe that tells the computer how to prioritize all the inputs to determine what to send to whom. The principal mandate of those algorithms: deliver what holds users attention (usually by making them feel good) so they will stay on the site for as long as possible (and see as many paid messages as possible).

The famous statement on the front page of the *New York Times,* "All the News That's Fit to Print," highlights the difference from social media. The *Times* describes its purpose as deciphering the fit from the unfit. Social media, on the other hand, have the purpose of delivering what makes the individual user feel good—something they agree with—or something that will make the user click to generate revenue.

When it comes to such clicks, "Content that evokes high-arousal positive (awe) or negative (anger or anxiety) emotions is more viral," a study in the *Journal of Marketing Research* found.[43] This means that when someone in Macedonia is compensated every time an American clicks on a headline, greed overcomes the search for truth.[44] When the best "clickbait" is that which triggers an emotional as opposed to a rational reaction, bold headlines unfettered by fact are the result.

For a century and a half, we have relied on the centralized nature of networks to outsource to a third-party editor the responsibility for making judgments on the source, veracity, and context of information. Now, in the medium used by the majority of Americans as their source of news, those editors have been replaced by machines running algorithms prioritized not for veracity but for velocity and economic optimization.

The effect of this is that, like the network itself, the role of information curator has been distributed outward.

Software algorithms got us into this situation; software algorithms should also be harnessed to get us out. Pushing the curatorial function outward means there is a need for public-interest algorithms to counter the economic-interest algorithms.

Where there can no longer be an expectation of objectivity, then there must be transparency. I have been privileged to work with Wael Ghonim on his shockingly simple idea: that the inputs to the social media algorithms must be opened to the public. A common occurrence in the software world is what are called "open APIs." An API—application programming interface—is what allows two software programs to interact with each other. An example is how Uber uses the information in Google Maps to create a taxi service.

Adoption of an open API by social media platforms would not mean revealing the "black box" secrets of the algorithm itself or exposing any personally identifiable information about users. But by opening up what goes into and out of the social media company's algorithm, third-party programmers could create public-interest algorithms to understand the effects of the social media distribution. Knowing who purchased ads or created posts, for instance, and combining that with information about reach, engagement, and demographics would allow a public-interest algorithm to assemble a picture of what is being spread about and to what kind of groups.

As important as understanding what is happening, is gaining that knowledge with computer speed. It takes only seconds for an ad or posting to spread throughout the world. Yet discovering that the distribution has occurred can take hours or days. Being able to track what's going in and coming out via a public-interest algorithm would permit the kind of curation for veracity that the platforms do not perform themselves.

Today, public-interest groups of all political stripes monitor the mainstream media. With a public-interest open API, these same groups could also build public-interest algorithms to accomplish the same result in social media. Hopefully, social media platforms would want to provide such transparency voluntarily. If they don't uniformly do so, then it will fall to government to facilitate such openness.

It is not as if we haven't seen this before. "They shamelessly print, at negligible cost, material which may, alas, inflame impressionable youths," wrote a Venetian scribe, bemoaning "degradation in the brothel of the printing press."[45] History's precedent has been clear: when the cost of information dissemination decreases, the nature of information necessarily evolves as well, bringing with it the need to develop new norms.

Digital Dividend/Digital Divide

In the African bush, a herdsman takes out his mobile phone and sends a text message to check the prices paid for cattle in nearby villages. At the same time, in Manhattan, an executive checks the app on his smart phone to see how his stocks are performing. After driving his cattle to the village, the herdsman receives compensation for his assets via a deposit into his mobile phone's account. Likewise, the proceeds from the executive's stock sale are loaded into his mobile-accessible ledger.

The new network is the Great Leveler. When an African herdsman and a

Manhattan executive engage in similar economic activities using a common platform, the network driving that platform is accomplishing something far beyond facilitating common experiences—it is creating common opportunity. For the first time in history, network capabilities available for one can be available for all.

At the dawn of the new millennium, the United Nations set an audacious goal to cut the world's poverty rate in half by 2015. The outcome was closer to a 60 percent reduction—one billion people emerging from extreme poverty.[46] New network connectivity—principally wireless—is universally cited as a critical driver of this reduction. According to the UN's International Telecommunications Union (ITU), more than 95 percent of the world's population is covered by a wireless signal.[47] That connectivity demarginalized the masses by opening capabilities the developed world had long taken for granted.

It is difficult to overstate the impact of the vast majority of the world's population sharing a basic level of common connectivity for the first time in history. The ability to make a call to check on a loved one, or to conduct business, or to summon emergency help—simple actions that were previously impossible—changes lives forever.

Connectivity is not just about making phone calls. In most of the world, access to water and electricity has always been limited because of the absence of a payment infrastructure to share the cost of the necessary facilities among users. But when 500 million households without electricity have a mobile connection (and the ability to use it for payments), the economic infrastructure for electrification exists. The same is true for the 700 million without access to clean water who have a cell phone.[48] While the vast majority of the world's population has never had a bank account with which to save and dispense funds, the system that allows for the payment for mobile phone service has also created the platform to make possible the construction and operation of almost everything else.

But while mobile voice and text have been transformational in their impact, access to the broadband internet remains aspirational to most. On average, eight in ten individuals in the developing world own a mobile phone, but only about three in ten have internet access.[49] Nor are we in the developed world immune from such realities when approximately 20 percent of the population is not online.[50]

The divide between those without internet access and those with access has the potential to exacerbate—rather than attack—economic inequality as those who have connectivity advance and those without fall farther behind. Whether innovative solutions are brought to bear, and how, is one of the critical decisions yet to be made about the implementation of our new networks.

Two principal factors—one economic and the other structural—govern access to broadband networks.

The cost of service to individuals remains a significant economic barrier. In the United States about half of households with incomes below $20,000 were not online in 2014.[51] One of the initiatives we developed at the FCC was to expand a program that had subsidized the cost of basic phone service for low-income households by applying the subsidy to broadband internet access.

One of the principal divides the FCC program was designed to attack was the "homework gap" that existed for low-income students.[52] Because the majority of U.S. public school students live in poverty, they are less likely to have internet access at home.[53] Even when schools provided a laptop, the students couldn't do their homework unless they went to a location with Wi-Fi, such as the local McDonald's.[54] Upgrading the government's traditional low-income support for phone service to include broadband internet was a step toward the kind of universal access to the network necessary for schoolwork, as well as for applying for a job and for virtually every other modern-day activity.[55]

Structurally, the cost of delivering broadband to areas of low population density limits the realization of its benefits. Running new fiber-optic lines to remote areas is costly. In the days of the phone monopoly, the costs of providing Theodore Vail's universal service to remote areas were buried in everyone's phone bill. Since the demise of the monopoly, the subsidies necessary to support high-cost areas have been more explicit through a fee on the bill that funds a program overseen by the FCC to subsidize rural phone companies. But even with these subsidies, wiring remote areas is too often a high-risk investment.

Wireless networks that overcame geography to provide voice service to the previously unconnected offer a similar potential for lower-cost broadband infrastructure. Fourth-generation (4G) wireless service provides low-level broadband internet connectivity, but it is expensive as its limited spectrum means it has limited capacity. Fifth-generation (5G) wireless technology offers faster broadband-like speed, but the spectrum limitation remains.

Hot on the heels of terrestrial wireless, however, are even newer network technologies with the potential for even lower costs. From circling drones and balloons to new constellations of satellites, the march of new network technology can once again reshape connectivity by eliminating terrestrial infrastructure, especially the physical improvements required to "backhaul" the data feed picked up by a network antenna to the rest of the network.

Proposals we dealt with at the FCC, for instance, included the 2019 planned launch of three new satellites with as much bandwidth as the 400 or so other satellites in the world combined.[56]

Further revolution is coming from the reduction in the cost of satellites. Previously, the cost of a new satellite was in the $200 million range; now thanks to new technology and assembly-line-like processes, it has fallen to approximately $1 million. Accompanying the reduction in satellite costs is the reduction in launch costs as commercial launch services replace NASA's monopoly. Taking advantage of these developments, two new constellations of 4,700 refrigerator-sized satellites in low earth orbit promise to deliver the internet everywhere, even to the most remote areas.[57]

New networks have produced spectacular dividends. The promise of continued innovation driving additional dividends is palpable. The question remains whether the fruits of innovation will be equally delivered, both among nations as well as within them.

Throughout history we have seen the power of connectivity to change lives. Whether our new connectivity fulfills its potential to expand opportunity for all, or simply deepens inequality, remains in contention.

Connecting Forward

Each of the chapters in this book closed with a story of how the particular technology it discussed resurfaced as a key part of a subsequent network revolution.

Johannes Gutenberg's envisioning information in its smallest parts returns in the language of the internet.

Charles Babbage's lament, "I wish to God these calculations had been executed by steam," opened the path to automated computing.

Samuel Morse's "flash of genius," while a triumph of narcissism over fact, nonetheless opened an era of electronically disembodied information.

And for the past half century, the modern extensions of these network revolutions have been churning to deliver the network forces that will determine our future.

Today, with a degree of anxiety, we observe the approach of newer effects of our network evolution. The takeaway from the previous network revolutions is that no one was prescient. Everyone, from the humblest individual to the greatest scientist to those whose innovations harnessed the new technology, was making it up as they went. Each was assembling a giant jigsaw puzzle without benefit of the picture on the top of the box.

History is told in seemingly clear-cut conclusions generated by a set of facts that appear self-obvious. Historians know how things turned out; contemporaries, like us, who are living through the upheaval have no such advantage. One thing we know from history, however, is that change is unruly, unstructured, and unpredictable. Decision-making in real time is messy and insecure.

I began this book by challenging the assumption that we are living through the most transformative period of history. We now turn to how technology has put us on a course capable of crossing that threshold.

Nine

Connecting Forward

Chapter 1 began with a discussion of Paul Baran's 1964 description of the move from a centralized network to a distributed network. After fifty years, Baran's concept has been widely implemented and is the structural revolution on which our future will operate.

A principle of this book is that it is never the primary network technology that is transformational but rather the secondary effects made possible by that technology. The previous chapter sampled some of the ongoing effects of Baran's network concepts. This chapter looks at how the distributed digital network underpins the four network-based forces that will define the future: a new generation of the web, artificial intelligence, distributed trust, and cybersecurity. Together they possess the potential to deliver a level of network-enabled commercial and cultural transformation that just might rival that of the last great network-driven revolution in the mid-nineteenth century.

The union of low-cost computing and ubiquitous connectivity across a low-cost distributed digital infrastructure has the capability to deliver us fully into the third great network-driven transformation. For as much as we talk about living in an information age, we have yet to experience a period to rival the Industrial Revolution. Our networks, for instance, while shaping many daily activities, have yet to have an industrial-age-level impact on the heart of the economy: the productivity of the creation of goods and services.

In his masterful *The Rise and Fall of American Growth,* economist Robert Gordon compared the growth in productivity of the years 1920–70 with the years that followed. The average annual growth in productivity per hour dropped from 2.8 percent in the mid-twentieth century to 1.6 percent as the digital era emerged.[1]

The early days of the internet's growth had a significant impact on economic productivity, but it was not sustained. Between 1996 and 2004, American productivity rode on the back of ubiquitous personal computing and the web to increase annually by an average of 3.1 percent.

But then productivity growth fell away. In the period 1970–2014 the average annual growth in productivity per hour was below that of the post–Civil War era. Certainly, the network continued to have an impact on our personal lives, but being able to order a pizza on a smart phone is a far cry from the expansion of efficiency in the production of goods and services at the heart of the economy.

The culprit was the network itself—and the new economic rewards structure it created.

With the introduction of the World Wide Web around 1990, the internet became usably accessible to mere mortals like you and me. While the early digital network had created the ability to access diverse databases of information, its use was limited to an internet priesthood and nerdy hobbyists. Tim Berners-Lee's creation of the web broke through those barriers by creating a common protocol for finding and displaying information from disparate databases across disparate networks in a simple request/response format.

The first iteration of the web (Web 1.0) gave us browsers with which to seamlessly search the world's information. Accompanying that capability, and driving its adoption, was the opportunity to sell advertising associated with the information. Adoption of the web coincided with and helped fuel the 1996–2004 decade of internet-driven productivity growth.

About a dozen years later, Web 2.0 democratized the network by allowing anyone to create and deliver information. It was the birth of social media. Consumer-facing, as opposed to productivity-enhancing, activities dominated the web. The economic model of selling to businesses a consumer's self-expressed interests took off to become the dominant economic model of Web 2.0.

Outside social media, the new business model did little to improve the production efficiency of converting inputs into outputs of goods and services—the basic measure of productivity. Despite all the innovations we have seen in our personal lives—from Facebook, to Netflix, to Waze—productivity growth slowed.[2]

There is a fundamental difference between Facebook or Netflix on your smart device and a revolution in the core production capabilities of the economy. "Economic growth since 1970 has been simultaneously dazzling and disappointing," Robert Gordon observed, because "advances since 1970 have tended to be channeled into a narrow sphere of human activity having to do with entertainment, communications, and the collection and processing of information. For the rest of what humans care about—food, clothing, shelter, transportation, health, and working conditions both inside and outside the home—progress has been slow."[3]

But the cavalry is on the way. The distributed digital network has become the infrastructure for game-changing and productivity-enhancing uses of the network. It all starts with a new iteration of the web itself.

Web 3.0

Creating Value by Orchestrating Intelligence

The new Web 3.0—called the semantic web by Berners-Lee—stands the web's traditional request-response structure on its head. As Moore's law enables microchips to be built into everything, wireless connections allow access to the functionality of N+1 number of chips and the intelligence they generate. Rather than the call-response of earlier versions of the web, in which existing information was discovered and displayed, Web 3.0 orchestrates intelligence to create something new.

Using the web to deliver a movie or a Facebook post is the *transportation* of information that has already been created. Web 3.0, in contrast, is the *orchestration* of a flood of intelligence from connected microchips. It is, for instance, the difference between a connected car and an autonomous vehicle.

The cars we drive today are wirelessly connected in the manner of earlier iterations of the web: there is an ongoing request-and-reply of information into and out of the automobile. Autonomous vehicles, in contrast, are full of microprocessors generating intelligence that must be orchestrated with that of other vehicles, road signs, weather sensors, and a myriad of other inputs. That orchestration creates a new product (the safe coexistence and cooperation of vehicles), and that product creates new productivity (more efficient uses of highways and roads).

Each autonomous vehicle is expected to produce 25 gigabytes of data per hour—the equivalent of a dozen HD movies.[4] Former Intel CEO Brian Krzanich estimated that in one day, a single car will generate about as much data as 3,000 people do in a similar period today.[5]

The autonomous vehicle is but one example of how the distributed network becomes a platform for new applications as the intelligence being produced by innumerable nodes throughout the network is connected so that it can be manipulated to create new products and drive productivity.

As we have seen, the earlier versions of the web produced brief efficiency gains. Putting computers with internet access throughout an enterprise increased productivity through the improved transmission of and access to information, but the next productivity jump was elusive. Web 3.0's semantic capabilities, however, promise continually increasing productivity improvements to accompany the exponential increase in intelligence generated by connected microchips.

The move from transporting preexisting information to orchestrating new intelligence to produce new products and services will redefine the economics of the network from *push* to *pull*. Thus far, the business model of the web has been dominated by pushing information to targeted users and selling that capability to advertisers. Web 3.0 redefines value creation as pulling the intelligence created by tens of billions of connected microchips so that it may be manipulated to create new products and capabilities.[6]

The first act of Web 3.0 has been dubbed the internet of things (IoT).

Consider, for instance, how connected microchips change the industrial process. A company such as Boeing, for instance, must orchestrate the activities of more than 28,000 suppliers and activities in seventy countries, not to mention products that are continually on the move. By including inexpensive microchips in component parts, Boeing can track the shipment of parts from suppliers for just-in-time arrival. Once on the assembly floor, wireless sensors read the whereabouts of the parts and instruct automated equipment. Then, when the finished product leaves the plant, the same kind of connected intelligence provides global, real-time tracking and tracing of the aircraft and its components.

It's not just in industrial settings where IoT enhances productivity. Sensors monitoring sunlight, humidity, and ground moisture help agricultural operations protect crops and maximize yield. When the produce is shipped to market, sensors track its movement to market, as well as how fast the produce is ripening. The latter information is especially instrumental to productivity as it can trigger cooler refrigeration or even delivery to a closer intermediary market.

IoT can even help reduce your water bill. In the typical public water system in the United States, 16 percent of the water put into the system never reaches the consumer.[7] Intelligent sensors placed strategically throughout the water system can identify and report leaks that otherwise would go unnoticed. A new product—real-time monitoring of the water system—thus results in increased productivity of the water system to save ratepayers money.

Whether in industrial, agricultural, or smart-city applications, the opening act of Web 3.0, the internet of things, creates a new information product: the real-time awareness of everything going on. The application of that information increases productivity.

The old economic model of the web had to be principally consumer-facing as it depended on advertising dollars. Making money with Web 3.0 will be different as intelligence becomes a raw material used to create new products that will in turn make those things more productive. The business question of Web 3.0 then becomes "What can I build?" as opposed to "What can I sell?"

Artificial Intelligence

Our Network Resembles Our Brains

As Charles Babbage struggled to explain his analytical engine in nineteenth-century terms, he described the first computer as "eating its own tail" because it based one calculation on the results of preceding calculations. It is a description that could also be applied to what we today label "artificial intelligence." And Babbage's mechanization of human reason raised, in the Victorian era, the same kinds of existential issues that have emerged around artificial intelligence today.

The term artificial intelligence (AI) first surfaced in a 1979 article by Stanford professor John McCarthy.[8] Since then it has been used and abused in popular culture and endless science fiction thrillers. Computer science legend Ray Kurzweil forecasted that by 2045, the ability of AI to continuously improve on itself would result in the "singularity"—machine-based superintelligence greater than human intelligence.[9] For the foreseeable future, however, our reality will be shaped by multiple levels of evolutionary computer intelligence that will creep into daily life.

One level of computer intelligence—often called machine learning—is the ability of machines to sift through great quantities of information to, Babbage-like, inform subsequent activities. Amazed that after you type only part of your search query, Google completes it for you? Pleased by how Amazon recommends books on topics of interest to you? Happy your radiograph is being read quickly and precisely? All of these are examples of "intelligent" machines accessing databases of previous searches, prior purchases, and earlier diagnoses to provide an answer.

The concept of an intelligent machine came to the nation's popular attention in 2011 when IBM's Watson computer beat two human champions on the TV game show *Jeopardy*. While it was referred to in shorthand as the computer "thinking," Watson really wasn't thinking. The format of the game show lent itself to a computer parlor trick where preloaded information produced answers that made the computer appear to be thinking.

The secret to *Jeopardy*, former champion Richard Cordray once told me, is a multistep preparation process. First, contestants must identify facts that lend themselves to the "What is . . ." answer format of the show. Then they must amass notebooks full of this information and learn those facts. It is an activity perfectly suited for an entry-level intelligent computer like the Watson of 2011. The basis of the computer's victory was that manmade algorithms followed Cordray's technique, but in instantly accessible digital code rather than stacks of notebooks (Watson's data included, for instance, the entire contents of *Wikipedia*). The algorithms would identify what was being asked, search the database for relevant information, and proffer an answer.

The next generation of intelligent computing programmed the computer to derive conclusions from data, not just spit out answers. "Conclusions" are different from "answers" in that the output is determined from the data being observed, rather than that which is programmed in. It is the difference, for instance, between the computer being told the Nile is the longest river in the world (an answer) and the computer searching geographic databases to report the answer (a conclusion).

As machines evolve from following specific instructions to the processing of diverse information, they begin to look like they are thinking—but they are not. What they are doing, however, is behaving in ways that mimic the neural network patterns of the human brain.

The brain is essentially an input-output system, just like a computer. There are 10 billion neurons that receive, process, and transmit inputs (for example, the stove is hot). The network connecting these neurons allows the brain to draw upon and assemble all these inputs to trigger an output response. Intelligent computing creates an artificial neural network of software and hardware to replicate the same process of harnessing diverse, distributed, but interconnected inputs to produce a conclusion.

As I stood in Boeing's new Composite Wing Center in Seattle—a building that could house twenty-five football fields—huge machines laid the strips of carbon-fiber-reinforced polymer used in the wings of Boeing's next-generation 777. To build the struts that run the length of the almost 118-foot-long wing, the machines painstakingly lay down 120 layers of carbon-fiber composite tape. The result is stronger yet lighter and more flexible than traditional metal struts.

My visit took place just after the new facility opened and before actual production began. From high above the floor, I watched half a dozen white-coated inspectors with magnifying glasses and flashlights poring over every millimeter of the gigantic strut checking for "gaps and [over]laps" in the most recent layer of composite tape. By feeding the "gaps and laps" into a database, the computer "learns" from its mistakes and "teaches" itself to eliminate the problems in subsequent runs.

The aircraft flying on these new wings might someday similarly be piloted by machine intelligence. The autopilots in use today are like the early Watson: loaded with instructions for what to do in specific circumstances. Confronted by something new, however, the autopilots default to human pilots to solve the unforeseen with their biological neural networks. Store all the learned experience of the humans in artificial neural network computers, however, and the automatic pilot could conceivably manage unanticipated circumstances in the same manner as human pilots.

At University College London, researchers have built an autopilot that calls on ten separate artificial neural networks, one for each different aircraft control (throttle, pitch, yaw, and so on). These systems then collect the digital records created by human pilots in simulators to determine how they respond to various circumstances, including the unpredictable. Using this database, the new autopilot builds a collection of information from which to draw in making decisions.

To utilize the huge amounts of data required for machine learning requires significant computing power. Accomplishing this has become a Gutenberg-like reassembly of diverse pieces of information, done at high speed. Interestingly, it draws on microchip designs created for video games. Graphic processing unit (GPU) chips were originally developed because to be realistic, video games required breaking large blocks of descriptive information into smaller pieces to be processed in parallel and reassembled on the fly into the finished product. It is exactly this parallel processing of large amounts of data that is necessary for machine learning.[10]

Machine AI, therefore, isn't really "thinking" but is rather a mathematical activity designed by cognitive individuals for the rapid mining and manipulation of information. True AI is a broader, more powerful concept whereby a machine becomes "smart" through its own intelligence. Thus far, it is more a concept than a reality (which is not saying it cannot become reality). Nonetheless, it has sparked its own debates. From White House workshops to the cautions of visionaries, the technical possibility and ethics of true AI remain a front-burner item. Long before we get to true AI, however, we will be forced to deal with the creeping effects of its early iterations.

During a meeting with the communications minister of Argentina about that country's plans for next-generation connectivity, the minister asked about the issue that has accompanied technological innovation throughout history and now circles around intelligent computing: what about the impact on jobs? Specifically, he referenced a 2013 study by two University of Oxford professors forecasting that 47 percent of the jobs in the United States were at risk of being replaced by machine intelligence.[11]

Such concerns have historically accompanied technological upheavals. The term "Luddite" made its way into the English language, for instance, when early-nineteenth-century English weavers protested the Industrial Revolution's automation of weaving looms. Characterized by the smashing of looms by the probably fictitious Ned Ludd, the movement galvanized around the fear that skilled textile workers would be replaced by machines that would require only unskilled operators.

Such automation produced the industrial age, a period when, indeed, sector-specific jobs were lost. However, the basic skills required in most jobs meant a worker could lose his job in one sector but still put his skills to work in another. The automation may have displaced workers from specific jobs, but it ended up being beneficial for the overall economy—and job creation—since the increased productivity led to lower prices, which drove up demand for products, and thus demand for workers.

Concern about job-killing automation resurfaced in the mid-twentieth century as computers began to come out of their sanctuaries. As far back as 1961 President Kennedy warned, "The challenge of the sixties is to maintain full employment at a time when automation is replacing men."[12] Two decades later, when PCs began appearing on desktops, the fear of being automated out of work returned.[13] Today, intelligent computing has reignited the issue yet again.

The challenge of an automated economy is not simply the replacement of lost jobs, it is the reallocation of workers and skills. The widespread adoption of bank automated teller machines (ATMs), for instance, substituted software and computer chips for a job done by humans. The number of bank tellers declined dramatically. This, however, had the effect of reducing the operating cost of each bank branch. As a result, banks opened more branches. The new branches didn't need as many tellers, but they did need sales and customer service personnel. The new job skills were different from those of a teller, but not that different—and they were harder to automate. As a result, automating tellers turned out to lead to an overall increase in bank employment.[14]

Intelligent machines also hold the potential to address some systematic problems, including, perversely, a potential shortage of workers. Over the next decade it is estimated that 3.5 million manufacturing jobs will open.[15] Yet the growth of the labor force available to fill those jobs is down by two-thirds owing to retiring baby boomers and lower birth rates.[16] Something is going to have to fill the void.

Automation could also mean the return of activities that for decades have been siphoned away by lower-wage countries. Why go to the expense and bother of assembling something offshore if it can be just as inexpensively assembled by intelligent machines domestically? Even if foreign countries adopt the technology themselves, they will still lose their cost advantage. Since foreign intelligent machines will presumably have the same capabilities as domestic machines, there is no economic gain from exporting the activity in the first place.[17] Operating the machines will likely represent a new occupation demanding new skills—and it will mean returning jobs.

It is important to be neither too rosy nor too dark about the challenges of machine learning and AI. It is doubtful the world will sit calmly by and not respond to the challenge of automation. Replacing 47 percent of jobs with machines may be a sustainable assumption in academia and in economic models, but this nightmare scenario can occur only if everyone sits on their hands. As machine intelligence will be implemented neither instantaneously nor in isolation, economic and political structures will have an opportunity to respond.

The ATM example is instructive. The jobs that were lost were rote activities that could be automated. The jobs that were created required creativity and the ability to solve problems and make decisions. Responding to intelligent machines isn't so much about creating new jobs as it is about new occupations.[18] Such a situation challenges us to reassess educational preparation to deprioritize skills that can be automated and reprioritize training for intellectual skills that can't be automated.

Automation also challenges us to revisit national policies built on industrial age assumptions. Workers' rights, for instance, have evolved as we've moved from the shop floor to the digital economy. Increased productivity became the rationale to cut the workweek by one third. Government policy and collective bargaining rebalanced the employer-employee relationship that industrialization had skewed. Without a doubt, similar adjustments will be necessary as we deal with the reality of machine intelligence.

But the greatest worker's right is the right to be prepared—both entering the job market and continuing in it. One study projects that 60 percent of the 3.5 million industrial jobs that would become available over the next decade will go unfilled because our education system isn't producing individuals with the requisite technical, computer, and problem-solving skills.[19] And when the CEO of AT&T tells employees to spend five to ten hours per week expanding their skills to "retool yourselves," the right to preparation takes on lifelong proportions.[20]

Machine intelligence will be essential to handling the tsunami of data from Web 3.0. Both the flood of product-producing data, and its automated applications will reprise the experience of upheaval and angst that we have seen technology-driven change create throughout this book. How we think about production, education, workforce allocation, and economic models must change. It was the pressure of earlier network revolutions that forced revolutionary education and labor policies that are today the accepted status quo. That experience is important to remember. Our technology may be new, but dealing with the effects of the change it creates is familiar.

Blockchain Trust

A Distributed Network Creates Distributed Trust

About forty years after Johannes Gutenberg's printing press, Venetian mathematician Luca Pacioli's book (see chapter 2) introduced double-entry bookkeeping to a world beyond the merchants of Venice. With his common principles of double-entry accounting, Pacioli established the basis for a coordinated banking system rooted in counterparty trust. A bank in a distant city knew it was safe to transfer money into a particular account based on a trusted relationship with the originating bank, a relationship that began with the knowledge both banks were keeping score the same way. The banks then built their businesses by charging a fee to use that trusted relationship.

All the world's transactions—not just cash transactions—have at their heart the need for trust between the parties. Unchanged over the centuries, trust was a hierarchical, hub-and-spoke system. Every time we pull out a credit card to make a purchase, for instance, that classic trust mechanism goes to work, and the banking intermediaries make money. The credit card company sells to the merchant the trust that the bill will be paid, while at the same time selling to the customer the trust that the merchant will honor the plastic. The credit card company's service functions, just like the railroad hub or telephone switchboard, are a centralized activity.

As Web 3.0 orchestrates the intelligence from tens of billions of microchips, counterparty trust becomes even more important. The validity of the source of the intelligence, and of the intelligence itself, will need to be affirmed in real time. The slow centralized trust mechanism will no longer be adequate to verify the request for and creation of the flood of information from N+1 microchips.

Creating the interparty trust required by any transaction—whether trading stocks, purchasing goods and services, or any other movement of an asset—has traditionally been a hierarchical activity built around a centralized ledger. The migration of activity to the edge of the network has created both the need for a more efficient trust-building mechanism—and also created the vehicle for its delivery.

Generically described as "blockchain," the distributed network has enabled a system of distributed ledgers. These distributed ledgers can do for internet transactions what the web did for the internet: bring a layer of simplicity and increased performance. The web made it possible to find and link to information in the vast morass of the internet. Blockchain creates similar links among ledgers that record value.

The earliest iteration of this distribution of trust was Bitcoin, a non-state-authorized pseudo-currency. Since any currency is a fiction backed up by the trust that the value is as represented, Bitcoin simply mimicked this online by moving the trust validation function out of a centralized governmental structure onto the distributed network. The pseudo-currency moves over the network in peer-to-peer transactions without the need for a centralized trust validator, but with validation nonetheless.

To accomplish this, each bitcoin has a complex ID number (called a hexadecimal code). Tracking this code replaces the institutional validator by what amounts to a decentralized super-spreadsheet accessible to anyone participating. The idea surfaced in a 2008 white paper signed by a still unknown person or group called "Satoshi Nakamoto." In place of handling financial transactions the same way as switching boxcars—that is, bringing everything to a central sorting and validation function—the white paper described a distributed ledger technology that follows the path of the distributed network to move activity outward to the multiplicity of points at the edge of the network.

The new distributed ledger technology became known as blockchain. What these distributed ledgers do is replace central databases of who owns what and who owes what with a network of duplicate databases holding the same information. Because they are networked, these databases are constantly updating each other with information on the latest transaction. It amounts to a synchronized global ledger of all executed transactions that is securely distributed across multiple physical locations. It tracks, verifies, and records all transactions; putting them on permanent display to inform all other transactions. Every transaction is a "block," and every block, when recorded in the ledger, creates a new asset-allocation reality by which the next block is measured—thus the term blockchain, a continuous chain of new blocks of information that are constantly updating reality.

When the distributed network is used to enable distributed ledgers, trust is created not by holding proprietary information but through a collaborative understanding in which everyone knows what everyone else owns, owes, and is doing.

Luca Pacioli turned accounting into math where capital is always a credit and cash is always a debit and the two must reconcile (assets = liabilities + equity). Pacioli's ledgers were authenticated by a mercantile officer, just as today's ledgers are authenticated by trust-checking auditors. Blockchain's giant collaborative ledger replaces those costly and time-consuming trust-checkers with algorithms that constantly validate the implementation of transactions.

Look at the credit card example again. When you present a piece of plastic, the merchant determines whether to trust that you will pay by running the card through a terminal linked to the credit card issuer. A simple lookup is performed by the credit card company to verify the user is creditworthy and returns a "trust it" message (or the dreaded "I'm sorry but your card has been declined"). For providing this service, the card company charges the merchant a percentage of the transaction.

But what if instead of going to the credit card company's proprietary database, the information request went to a networked set of shared ledgers? In this instance, the algorithm created by the merchant's inquiry would cause the distributed spreadsheets to update, deduct the appropriate amount from the customer's account, and credit the merchant's account (or alternatively decline the transaction). A proprietary information-hoarding activity would be replaced by collaborative sharing of information.

Because it is handled electronically on a network with declining marginal costs, the expense of blockchain validation plummets. By moving from a proprietary to a collaborative ledger, the extraction of fees to use a centralized database also declines. Just as the distributed delivery of a phone call replaced the high cost of a centralized system for communications, so does the use of a distributed ledger replace the high cost of conducting a transaction using a dedicated trust agent. At the same time, fraud and dispute resolution are reduced as the super-spreadsheet knows *all* the customer and merchant information, not just the information that might be in one company's records.

While blockchain's first application was Bitcoin, the technology can record and report on any kind of transaction. Any item that can be tracked by a ledger can be secured with a distributed ledger. Because everything has a supply chain coming from somewhere and going somewhere else, everything can be tracked.

Counterfeit drugs, for instance, are a life-and-death threat when unscrupulous middlemen violate their position of trust and substitute look-alike pills for the real thing. With blockchain's ledger tracking and recording of every transfer, however, the compromise can be caught. Similarly, the serial numbers etched into diamonds can be easily and openly tracked and available to all, making them harder to resell and less likely to be stolen, and exposing illegal extraction practices. And for assets where provenance of the object is essential to its value, such as in the art business, tracking the item on a blockchain ledger guards against both theft and forgeries. Distributed ledger technology can secure anything that can be entered into such a ledger—from the serial number of an expensive piece of electronics to the tracking number on livestock.

Blockchain also is a potential solution to the problems created by the electronic network's peer-to-peer redistribution of assets such as music. Before the peer-to-peer transaction capabilities of the internet, record labels provided the trust function to protect copyrights by controlling distribution and preventing copying. The arrival of digital music storage and peer-to-peer networking bypassed that function, and with it payments to artists and composers. However, if music were delivered peer to peer using blockchain, a file on the ledger would exist for each piece of music. Each file could then even have its own unique rules (for instance, "play once for free, thereafter debit the player's account and credit the artist's").[21] And blockchain is tailor-made for micropayments for each play, something that has been impossible due to the high cost of centralized settlement systems.

Just possibly, the same kind of distributed ledger model could help each of us recover control of our private information. Today, huge centralized databases suck up personal information and sell it. Just as blockchain can create a micropayments structure for music, it can create individual micropayments every time a piece of our personal information is used. Platform companies such as Google and Facebook make huge profits by brokering the sale of such personal information. Blockchain can create a means for individuals to control that information, determine what to make available, and even receive payment for it. If blockchain can eliminate centralized credit card companies' brokering of personal credit information, it could do the same for all other information and put consumers back in control of their privacy.

As the distributed network expands the flow of data and the universe of transactions, a fast, low-cost means for validating those transactions becomes essential. Blockchain takes Pacioli's principle of a standardized protocol providing trust and disperses it throughout a distributed network to expand the provision of trust necessary for a successful transaction.

Cyber Vulnerability

When Everything Is Connected, Everything Is Vulnerable

From the beginning of time, network pathways have been avenues of attack. Primitive cultures mounted attacks by following animal paths. Alexander and Caesar conquered the world using roads and waterways; "all roads lead to Rome" for a reason. Britannia ruled the waves to keep open the trade routes of the empire. And in the twentieth century the nations of the world simultaneously used land, sea, and air pathways to deliver the bloodiest decade in history.

The digital pathways of the twenty-first century are no exception. Therefore, we should not be surprised that the new digital networks have become the new pathways for attack.

Exploitation of the new network can vary from criminal activity to intelligence collection and acts of war. The openness, ease of access to, and interconnection of a multiplicity of digital networks can aid and abet these diverse exploitations.

As the nature of what is connected has changed, the opportunity and incentive for cyberattacks have increased. The early digital networks simply connected computing machines; the security threat was the mischief of hackers or vandals. As data storage became both costless and essential, customized attacks on corporate and governmental databases became a tool of espionage, blackmail, and exploitation. Then, as networks began connecting people, mobile devices, social networks, and personally identifiable services became a target-rich environment for criminals and nation-states. Finally, the internet of things' tens of billions of connected microchips has become the addition of tens of billions of new attack vectors.

Once a week while I was FCC chairman, I would go to the agency's

SCIF (Sensitive Compartmented Information Facility) for an intelligence briefing about activities occurring on the networks that interconnect our nation and the world. The conclusion was clear: when everything is connected, then everything is vulnerable.

The training and software necessary to mount a cyberattack are easily available on the network itself. Without any difficulty, it is possible to purchase software that when pointed at a target automatically starts hacking into it. On the open market, zero-day vulnerabilities (an exploitable flaw in a piece of software) can be purchased to attack the software that runs servers, routers, and the devices that connect to them. It is even possible to use a Google-like web search to identify IoT devices so that they can be hacked and repurposed to become a cyberattacker controlled from afar, with the legitimate user none the wiser.

The internet is like the old New England commons: a shared space accessible to all for the greater good of all. Yet history has also shown the tragedy of the commons: individual action without collective responsibility tend to result in a muddy, overgrazed commons of diminished value. The commons of the internet invites similar abuse, amplified not just by greed but by a new definition of warfare.

This is the consequence of internet design decisions made decades ago. "If we had waited to build a perfectly secure network, there never would have been an internet," Bob Kahn, one of the fathers of the internet, once told me.

Because the original design of the internet assumed mutual trust among users, it is speckled with systematic security problems. The application boom triggered by the web expanded those weaknesses with an economic focus built on speed to market, not secure design.

We live with the hydra-headed consequences of those decisions. There was neither malfeasance nor misfeasance, but the result has been that our infrastructure carries both legitimate and harmful traffic. By one estimate, for instance, malicious botnets (computers controlled without their owner's knowledge) account for 30 percent of internet traffic.[22]

A key challenge in dealing with cybersecurity is the broad and diverse range of how and where the network can be exploited.

Cyberattacks are a threat to infrastructure. A cyberattack on Ukraine's power grid in 2015 left 700,000 people without electricity for hours.[23] In 2013, Iranian hackers attacked a dam outside New York City.[24] In 2016, a U.S. court convicted a Russian of attacks that caused more than $169 million in losses to 3,700 financial institutions.[25]

Cyberattacks enable information warfare. During the 2016 U.S. presidential election, intelligence agencies identified Russian-instigated hacks. By collecting and releasing private information from individuals and political institutions, the attackers shaped the discussion of issues in the election, and perhaps its outcome.

Cyberintelligence collections purloin intellectual property. General Keith Alexander, the former director of the National Security Agency, described attacks by foreign entities on U.S. intellectual property as "the greatest transfer of wealth in history."[26] Once inside a network, with the push of a key it is possible to download in a matter of seconds information that, had it been stolen in hard copy, would fill multiple eighteen-wheel trailers. Such attacks aren't limited to foreign attackers—they can be as American as baseball. In 2015, the scouting director of the St. Louis Cardinals was convicted of hacking into the Houston Astros' computers to collect scouting and player personnel information.[27]

As the internet went mobile it became personal, opening up the opportunity to exploit individuals. The ability to reach your contacts, calendar, and all your activity on the internet is only the beginning of mobile vulnerability. As mobile phones are typically synched to your PC, they are an inviting path into your desktop, and from there into everything else in your digital life. Addicted to that connected watch on your wrist? Think of it as yet another attack vector where infectious software can be implanted by proximity to a bad guy's network (think of the "other networks" that show up on your phone but you ignore), then jump to your PC and from there to a free ride anywhere on the internet. And infected mobile devices can literally walk right past perimeter-protecting security to attack a target.

As the IoT grows to connect tens of billions of microchips, encompassing everything from security cameras to light bulbs, it opens new avenues of attack. By one estimate, it takes only six minutes from the time a device is connected to the internet for it to be discovered and infected so that, without its owner knowing, it is under the command of someone else.[28]

We have failed to collectively address the cybersecurity problem. Both market forces and government have been insufficient and ineffective. Networks can increase security through more secure protocols, monitoring, and filtering. Device manufacturers can make security—including supply-chain reliability—a forethought rather than an afterthought. Government could step up with thoughtful and comprehensive legislated policies rather than relying on administrative interpretations of pre-digital-age statutes.

"Anytime you have a dependency on the internet, we're gonna be playing catch-up in reaction to defending our networks," former director of national intelligence James Clapper warned.[29] That we have such a dependency is a given. The necessity of getting beyond catchup continues to grow.[30]

We survived the nuclear threat and the scourge of chemical weapons through policies of containment. But containment is the opposite of the distributed force of the internet.

Cyber vulnerabilities are like the classic science fiction tale where the helpful robot turns avenging attacker. Any computer connecting to any other computer in the world—the essence of the internet—means it is possible to compromise, take control of, and exploit any computer-connected systems—and do so at scale. "The Cold War is over," wrote the *Washington Post*'s David Ignatius. "The cyber war has begun."[31]

Epilogue

There really can be no close to this technological travelogue. What's more, the trip continues at an accelerating pace. That is what makes an appreciation of the relationship of tomorrow and yesterday so darn important.

One of the late Alvin Toffler's delightful practices was the occasional convening of an evening of fascinating individuals, each with his or her perspective on technology and its impact. In an earlier era it might have been described as a "salon." It was at one such event, on a spring evening in Washington, D.C., that, after more than an hour of spirited discussion, I observed how our conversation had not touched on any of the topics with which the participants were involved. Instead, we had been pursuing philosophical and theological themes.

Some of the participants that evening were advancing new technologies, others were manipulating the human genome, yet others were deep in national security. But when the participants hung up their lab coats and turned off their computers they, like everyone else, were searching for anchors in a storm of change that, in many ways, they were helping to create.

"Isn't it interesting," I commented, "that a room full of change creators is searching for truths to bring perspective to their change?" There were only two places we could turn for this shelter, I suggested: faith and history. Faith has always provided the perspective that "there's something bigger than me," and history is the collected experiences of people like us as they dealt with their own (surprisingly similar) challenges. Our faith, in fact, is inseparable from our history. We study the ancient scriptures for meaning in our modern lives because they tell the stories that provide insight into the universal human condition.

As the collected stories of the human journey, history offers the fundamental lesson that the challenges we face today aren't unique. No matter how much we flatter ourselves with self-absorption, we are but the continuation of the human saga.

I hope that notion remains this book's takeaway: that our networked revolution is technologically iterative and sociologically similar to the previous network revolutions of history. The stories that brought us to this point and are defining our future are incredibly interesting. What is special for us is that we inhabit a time when the combined forces of history and technology converge to, yet again, challenge us with change.

We know the stories that led us to this moment. We know how the actions of those who dealt with history's changes created our today. Now we are in a historic moment of our own, and it's our turn to guide how new technology determines the future.

Notes

Prologue

1. Perhaps because "breaking things" was a bit too aggressive for a company undergoing public scrutiny, Facebook has changed its motto to "Move fast with stable infrastructure."
2. "Digital Transformation Is Racing Ahead and No Industry Is Immune," *Harvard Business Review,* July 19, 2017.
3. Benjamin Mullin, "The Associated Press Will Use Automated Writing to Cover the Minor Leagues," Poynter.org, June 30, 2016.
4. In 1983, 91.8 percent of people twenty to twenty-four years old had a driver's license; in 2014 the figure had fallen to 76.7 percent. "Recent Decreases in the Proportion of Persons with a Driver's License across All Age Groups," University of Michigan Transportation Research Institute, updated April 3, 2018 (www.umich.edu/~umtriswt).
5. "Smart Tampon? The Internet of Every Single Thing Must Be Stopped," *Wall Street Journal,* May 25, 2016.
6. This wonderfully descriptive alliteration was first suggested by my friend Blair Levin.

7. Walter A. McDougall, *Throes of Democracy* (New York: HarperCollins, 2008), p. 106.
8. Rudi Volti, *Society and Technological Change* (New York: St. Martin's Press, 1955), p. 17.
9. David Sarno, "Murdoch Accuses Google of News 'Theft,'" *Los Angeles Times,* December 2, 2009.
10. Vicar of Croydon, preaching at St. Paul's Cross, cited in Gertrude Burford Rawlings, *The Story of Books* (New York: D. Appleton and Co., 1901).
11. Henry David Thoreau, *Walden* (1854; Princeton University Press, 1971), p. 60.
12. Allbutt Clifford, "Nervous Diseases and Modern Life," *Contemporary Review* (London) 67 (1895), p. 214.
13. Rosyln Layton, "Does the Internet Create or Destroy Jobs? A Snapshot from the Global Debate on Digitally Enabled Employment," *AEIdeas* (blog), American Enterprise Institute, December 29, 2014.

Chapter 1

1. I was fortunate to call Paul Baran a friend. He was an exceedingly humble man. The concept of packet switching was first theorized by Leonard Kleinrock of MIT in 1961. J. C. R. Licklider, also of MIT, had envisioned a "galactic network" of interacting computers in 1962. In parallel with Baran's work (and unbeknownst to each other), Donald Davies and Roger Scantlebury were doing similar work in the United Kingdom; it was this work that led to the term "packet switching," a more elegant description than Baran's "hot potato."
2. Let's stipulate at the outset that numerous networks have affected the patterns of our lives. Water and electricity redefined daily life. Highways and airways recast physical movement. The focus of this book is the networks that have inexorably led to the marriage of computing and communications.
3. It is possible to consider everything from a writing stylus to the wheel as some form of technology. The term "technology-based network," however, applies to a network enabled by mechanical (or electromechanical) capabilities.
4. Manuscripts were moved from one scriptorium to another during the Middle Ages, but that did not constitute a "network" in the sense that the information was broadly exposed and disseminated.
5. A very few exceptions relied on sound or sight to move information faster than human travel. Drum signals, smoke signals, or flashes of light could all

signal over distance but were for one reason or another constrained in their application.

6. See Tom Standage, *The Victorian Internet: The Remarkable Story of the Telegraph and the Nineteenth Century's On-line Pioneers* (New York: Berkeley Books, 1999).
7. The term "electronic" is used as defined in *Webster's New World Dictionary*, 3rd College Edition (New York: Simon & Schuster, 1988): "1. of electrons, 2. operating, produced, or done by the action of electrons or by devices dependent on such action."
8. Michael Clapham, *Printing: A History of Technology from the Renaissance to the Industrial Revolution*, ed. Charles Singer, E. J. Holmyard, A. R. Hall, and Trevor Williams, 4 vols. (Oxford University Press, 1957), p. 37. Cited in Elizabeth L. Eisenstein, *The Printing Press as an Agent of Change* (Cambridge University Press, 1979), p. 45.
9. Cited in W. Brian Arthur, *The Nature of Technology* (New York: Free Press, 2009), p. 17. Arthur calls this "combinatorial evolution" and applies it to technology.
10. Interview by Judy O'Neill, March 5, 1990, Menlo Park, Calif. See "An Interview with Paul Baran," OH 182, Charles Babbage Institute, University of Minnesota, Minneapolis (https://conservancy.umn.edu/handle/11299/107101).
11. Angus Maddison, "World Population, GDP and Per Capita GDP, 1–2003 AD," chart, Groningen Growth and Development Centre.
12. "Speed of Animals: Horse, *Equus ferus caballus*" (http://www.speedofanimals.com/animals/horse).
13. The calculation assumes the typical letter consisted of around 2,200 words and the typical English word contains, on average, 4.79 characters, as calculated by Peter Norvieg at Google. See Ed Vielmetti, "What Is the Average Number of Letters for an English Word?," Quora.com, February 11, 2015. If each character requires the standard 8 bits of data, then the letter would contain approximately 84,000 bits (2,200 × 4.79 × 8 = 84,304). If it took a month for the letter to reach its destination, then the effective throughput would be 0.03 bits per second (84,304 bits ÷ 30.4 days/mo. = 2,773 bits/day throughput ÷ 24 hours/day = 116 bits per hour ÷ 60 min./hour = 1.9 bits per min. ÷ 60 seconds/min. = 0.03 bits per second). See Principia Cybernetica (http://pespmc1.vub.ac.be/TECACCEL.html).
14. John F. Stover, *The Routledge Historical Atlas of the American Railroads* (New York: Routledge, 1999), p. 21.
15. Menahem Blondheim, *News over the Wires: The Telegraph and the Flow*

of Public Information in America, 1844–1897 (Harvard University Press, 1994), p. 17 (chart).

16. Standage, *Victorian Internet*, p. 57.
17. It takes a little over two seconds to tap in the Morse code for one average-size 8-bit character, thus producing a transmission rate that is approximately 3 bits per second.
18. Wireless speeds remain significantly slower than wired speeds, but increasingly at broadband speeds the value of the connection is enhanced by its ubiquity.
19. Increasingly, the decision making of individual hubs takes the user to a centralized corporate hub such as Google or Facebook, where unique decision making is replaced by algorithms programmed to hold users' interest to keep them online so they will see more paid messages. While this diminishes the individual's role determining in-out activity, the network continues to vest that power at the edge.

Chapter 2

1. For example, in France the twelfth-century cleric Peter Waldo tried to launch a reform movement. In fourteenth-century England John Wycliffe led a similar charge.
2. Stephen J. Nichols, *The Reformation: How a Monk and a Mallet Changed the World* (Wheaton, Ill.: Crossway, 2007), p. 30.
3. Elizabeth Eisenstein, *The Printing Revolution in Early Modern Europe* (Cambridge University Press, 1983), p. 151.
4. Luther at times hid behind unnamed friends who, he said, had given his works to printers. It was a thin veil.
5. Nicole Howard, *The Book: The Life Story of a Technology* (Johns Hopkins University Press, 2009), p. 58.
6. In 1508 the printer Johannes Rhau-Grunenberg was enticed to move his operations to Wittenberg. The new university in the town required printed texts. The founding of Wittenberg University at the turn of the century, in fact, had created a speculative bubble in the printing business. Five printers set up shop in the small German town, thus creating a supply in excess of demand. The Wittenberg printing bubble burst, but the basic level of demand remained, and Herr Rhau-Grunenberg was recruited to meet the needs of a more rational market. See Andrew Pettegree, *The Book in the Renaissance* (Yale University Press, 2010), p. 92.
7. Ibid.

8. Statement to his brother, Giuliano, as quoted in William Samuel Lilly, *The Claims of Christianity* (1894), p. 19.
9. Pettegree, *Book in the Renaissance*, p. 93.
10. Nichols, *Reformation*, p. 29.
11. Pettegree, *Book in the Renaissance*, p. 94.
12. Recent scholarship has begun to ask whether the story of tacking the theses to the church door was apocryphal. Irrespective of such a debate, there is no doubt about the role played by the printing press in the propagation of Luther's ideas. In fact, the possible absence of a public posting from which the theses could be copied even suggests a more complicit arrangement between the monk and the printers.
13. Lucien Febvre and Henri-Jean Martin, *The Coming of the Book* (London: Verso, 1997), p. 290.
14. Rossiter Johnson, ed., *The Great Events by Famous Historians*, vol. 8, *The Later Renaissance: From Gutenberg to the Reformation* (London: Aeterna Publishing, 2010), p. 18.
15. There is an ongoing debate over whether the printing press created the Reformation. While it is an interesting historical exercise, it does not add to our understanding here. The fact of the matter was that Luther's writing, as disseminated by the network of printers, fell on a fertile field waiting to be watered.
16. Pettegree, *Book in the Renaissance*, p. 95.
17. Febvre and Martin, *Coming of the Book*, p. 294.
18. John Man, *Gutenberg: How One Man Remade the World with Words* (Hoboken, N.J.: John Wiley & Sons, 2002), p. 273.
19. Ibid., p. 276.
20. Gutenberg was not alone in this quest; history records the efforts of others, such as Coster of Haarlem in Holland (and probably forgets the role of many others). Clearly, however, Gutenberg was the first large-scale implementation and (thanks to his legal problems) the best documented.
21. Albert Kapr, *Johannes Gutenberg, The Man and His Invention*, translated from the German by Douglas Martin (Aldershot, U.K.: Scolar Press, 1996), pp. 75–81.
22. Diana Childress, *Johannes Gutenberg and the Printing Press* (Minneapolis: Twenty-First Century Books, 2008), p. 35.
23. Man, *Gutenberg*, p. 76.
24. Precise terminology might dictate that the term "printing" actually applied to the woodblocks, while Gutenberg developed "typography." This text, however, sticks with the common description "printing" to refer to the process of transferring ink from raised type to a page.

25. Others had similar ideas about individual letters, but typically they reverted to casting the assembled collection *en bloc* so that it resembled a print block. See Febvre and Martin, *Coming of the Book*, p. 31; Man, *Gutenberg*, pp. 116–19.
26. Nicole Howard, *The Book* (Johns Hopkins University Press, 2009), p. 37.
27. Febvre and Martin, *Coming of the Book*, p. 31.
28. Ibid., p. 35.
29. Man, *Gutenberg*, p. 136.
30. Childress, *Johannes Gutenberg*, p. 56.
31. The resulting mixture was 5 percent tin, 80 percent lead, and 15 percent antimony.
32. There have been discovered what appear to be scraps of earlier printing runs, possibly trials, that were reused and pressed to make up hardcover bindings.
33. James Thorpe, *The Gutenberg Bible* (San Marino, Calif.: Huntington Library Press, 1999), p. 26.
34. Ibid., p. 29.
35. Eisenstein, *Printing Press*, p. 49; Febvre and Martin, *Coming of the Book*, p. 28, Johnson, *Later Renaissance*, p. 28. If true, this experience did not apparently discourage Fust from returning to Paris with his products. Historical evidence suggests it was in Paris while selling books in 1466 that he succumbed to the plague (Eisenstein, *Printing Press*, p. 50 n.).
36. Jane Gleeson-White, *Double Entry: How the Merchants of Venice Created Modern Finance* (New York: W. W. Norton, 2011), p. 70.
37. The Renaissance was under way in Northern Italy at the time of Gutenberg's discovery, and thus printing cannot have "caused" its occurrence, but printing certainly aided its expansion.
38. Eisenstein, *Printing Revolution*, p. 13.
39. Erwin Panofsky, "Art, Science, Genius: Notes on the 'Renaissance-Dämmerung,'" in *The Renaissance: Six Essays*, ed. Wallace K. Ferguson (New York: Henry Holt, 1962), p. 128, cited in Eisenstein, *Printing Revolution*, p. 140.
40. See Gleeson-White, *Double Entry*.
41. Toby Lester, *The Fourth Part of the World* (New York: Free Press, 2009), pp. 250–52.
42. Steven Johnson, *How We Got to Now: Six Innovations That Made the Modern World* (New York: Riverhead Books, 2014), p. 4.
43. See Tobias Dantzig, *Number: The Language of Science* (New York: Plume, 2007).

44. Marshall McLuhan, *The Gutenberg Galaxy* (University of Toronto Press, 1962), p. 228.
45. Jonathan Sallet, "Technology and Democracy: Dynamic Change and Competition," Twelfth Annual Aspen Institute Conference on Telecommunications Policy, Competition, Innovation and Investment in Telecommunications, August 11, 1997.
46. Eisenstein, *Printing Revolution*, p. 83.
47. Benson Bobrick, *Wide as the Waters* (New York: Penguin Books, 2001), p. 86.
48. Man, *Gutenberg*, p. 91.
49. Febvre and Martin, *Coming of the Book*, pp. 244–45.
50. Gertrude Burford Rawlings, *The Story of Books* (New York: D. Appleton and Co., 1901).
51. Ann Blair, "Reading Strategies for Coping with Information Overload ca. 1550–1700," *Journal of the History of Ideas* 64, no. 1 (2003), pp. 11–28.
52. Ibid.
53. James J. O'Donnell, "The Pragmatics of the New: Trithemius, McLuhan, Cassiodorus," in *The Future of the Book*, ed. G. Nunberg (University of California Press, 1996).
54. Paul Johnson, *The Renaissance: A Short History* (New York: Modern Library, 2000), p. 16.
55. McLuhan, *Gutenberg Galaxy*, p. 124.
56. Smithsonian Institution, *The Smithsonian Book of Books* (Washington, D.C.: Smithsonian Books, 1992), p. 122.
57. For a discussion of the debate over the origin of Morse code, see Carleton Mabee, *The American Leonardo: A Life of Samuel F. B. Morse*, rev. ed. (Fleischmanns, N.Y.: Purple Mountain Press, 2000), pp. 201–06.
58. Kenneth Silverman, *Lightning Man: The Accursed Life of Samuel F. B. Morse* (New York: Alfred A. Knopf, 2003), pp. 164–65.

Chapter 3

1. John F. Stover, *The Routledge Historical Atlas of the American Railroads* (New York: Routledge, 1999), p. 18.
2. William F. Baringer, *Lincoln Day by Day: A Chronology 1809–1865*, vol. 2, *1849–60* (Dayton, Ohio: Morningside Bookshop, 1991), p. 10.
3. George Rogers Taylor, *The Transportation Revolution, 1815–1860* (Armonk, N.Y.: M. E. Sharpe, 1951), chart, p. 79.

4. Gordon Wood, *Empire of Liberty* (Oxford University Press, 2009), p. 55.
5. Clifford F. Thies, "Development of the American Railroad Network during the Early 19th Century: Private Versus Public Enterprise," Independent Institute Working Paper 42, October 2001 (http://www.cato.org/pubs/journal/cj22n2/cj22n2-4.pdf).
6. Ibid.
7. The economic range of a bushel of wheat has been estimated to be about 200 miles, Arthur T. Hadley, *Railroad Transportation,* 1886, cited in Sarah H. Gordon, *Passage to Union: How Railroads Transformed American Life, 1829–1929* (Chicago: Ivan R Dee, 1997), p. 149. A bushel of wheat weighs 60 pounds and an acre could yield 25 bushels. Thus, a 100-acre farm would produce 75 tons of wheat. See Walter A. McDougall, *Throes of Democracy* (New York: HarperCollins, 2008), p. 128.
8. Nicholas Faith, *The World the Railways Made* (New York: Carroll & Graf, 1990), p. 115.
9. William Cronon, *Nature's Metropolis: Chicago and the Great West* (New York: W. W. Norton, 1991), p. 23.
10. Albro Martin, *Railroads Triumphant: The Growth, Rejection & Rebirth of a Vital American Force* (Oxford University Press, 1992), p. 166.
11. Patrick E. McLear, "The Galena and Chicago Union Railroad: A Symbol of Chicago's Economic Maturity," *Journal of the Illinois State Historical Society* 73, no. 1 (Spring 1980), pp. 17–26.
12. Stewart H. Holbrook, *The Story of American Railroads* (New York: Crown, 1947), p. 134.
13. Donald L. Miller, *City of the Century: The Epic of Chicago and the Making of America* (New York: Simon & Schuster, 1996), p. 95.
14. McLear, "Galena and Chicago Union Railroad."
15. Martin, *Railroads Triumphant*, p. 166.
16. Ibid., p. 82
17. Stover, *Routledge Historical Atlas*, p. 23.
18. It wasn't until 1874 that St. Louis finally agreed to a rail bridge. By that time Chicago had long since become ensconced as the rail center of the West.
19. Holbrook, *Story of American Railroads*, p. 101.
20. Bessie Louise Pierce, *A History of Chicago,* vol. 2 (New York: Alfred A. Knopf, 1940), p. 57.
21. Richard C. Overton, *Burlington West* (Harvard University Press, 1941), p. 30; George H. Douglas, *Rail City: Chicago USA* (Berkeley, Calif.: Howell-North Books, 1981), p. 41.

22. Maury Klein, *Unfinished Business: The Railroad in American Life* (University Press of New England, 1994), p. 10.
23. Martin, *Railroads Triumphant*, p. 167.
24. Christian Wolmar, *Blood, Iron, and Gold* (New York: PublicAffairs, 2010), p. 4.
25. A. H. Wickens, *The Dynamics of Railway Vehicles—From Stephenson to Carter,* Proceedings of the Institution of Mechanical Engineers, Part F. *Journal of Rail and Rapid Transit* 212 (1998), p. 209.
26. Simon Winchester, *The Men Who United the States* (New York: HarperCollins, 2013), p. 248.
27. The developer of this steam-powered suction, Thomas Savery, wrote, "My engine at 60, 70, or 80 feet raises a full bore of water with much ease." In practical application it was limited to less than that. Regardless, a 60–80-foot-deep mine leaves a lot of coal below it. Carl T. Lira, *Introductory Chemical Engineering Thermodynamics* (www.egr.msu.edu/~lira/supp/steam/savery.htm).
28. By accident, one day Newcomen discovered that letting water into the cylinder cooled the steam even faster and created even greater pulling power. But there remained the inefficiency of having to reheat a now-cooled container.
29. Fred Dibnah and David Hall, *Age of Steam* (London: BBC Worldwide, 2003), p. 44.
30. *Railroad History,* National Railroad Museum, Green Bay, Wis.
31. National Museum of Wales, *Richard Trevithick's Steam Locomotive*, 2008.
32. A few years later, on Christmas Eve 1803, the people of the area were startled to see a Trevithick steam engine sitting atop a carriage moving through the streets without assistance. Unfortunately, the contraption wouldn't make it past the holidays. Celebrating at the local pub, Trevithick's assistants forgot to extinguish the fire in the boiler; the pressure grew, and the engine exploded. William Rosen, *The Most Powerful Idea in the World* (New York: Random House, 2010), p. 290.
33. The loss in pressure was equivalent to the atmospheric pressure (14.7 lbs./sq. in. at sea level). This insight came from a leading scientist of the time, Davies Gilbert, with whom Trevithick had become friendly and from whom he sought advice. Anthony Burton, *Richard Trevithick: Giant of Steam* (London: Aurum Press, 2000), p. 59.
34. Rosen, *Most Powerful Idea*, p. 296.
35. Even before Trevithick, an American, Oliver Evans, had developed a similar concept (the 30° calculation cited, in fact, was Evans's). While Evans's

engine was demonstrated on water, it was never used on land (where the friction is much higher). Evans published his findings and even sent them with a friend to England "to be shown to the steam engineers there." Whether Richard Trevithick ever saw them is unknown; however, his concept is the same as Evans's. See Rosen, *Most Powerful Idea*, pp. 287–88.

36. Wolmar, *Blood, Iron, and Gold*, p. 6.
37. Rosen, *Most Powerful Idea*, p. 301.
38. George H. Douglas, *All Aboard! The Railroad in American Life* (New York: Smithmark Publishers, 1996), p. 17.
39. *The Observer* newspaper, cited in *Age of Steam* (www.railcentre.co.uk).
40. Wolmar, *Blood, Iron, and Gold*, p. 10.
41. The opening day of the Liverpool and Manchester Railway was also the occasion of the first bystander fatality. With the great and the good assembled, Liverpool MP William Huskisson failed to cross the tracks swiftly enough and was struck by the speeding locomotive *Rocket. Rocket* made a heroic run at speeds up to thirty-six miles per hour to take its victim to the nearest hospital. Huskisson did not survive. Rosen, *Most Powerful Idea*, p. 309.
42. Steel is an alloy that includes iron, a mineral. While early instances of horse-drawn wagons had rolled on iron rails, the strength and durability of steel in time made it the common substance of rails.
43. Jacques Barzun, *From Dawn to Decadence* (New York: HarperCollins, 2000), p. 539.
44. Douglas, *All Aboard!*, pp. 20–21; Taylor, *Transportation Revolution,* p. 77.
45. Douglas, *All Aboard!*, p. 22.
46. Cooper, a man with little formal education, would become wealthy from his industrial activities, move to New York City, and found Cooper Union, where Abraham Lincoln would deliver his famous 1860 "right makes might" speech.
47. John Steele Gordon, *An Empire of Wealth* (New York: HarperPerennial), 2004, p. 150.
48. Douglas, *All Aboard!*, p. 23.
49. Holbrook, *Story of American Railroads*, p. 24; Douglas, *All Aboard!*, p. 223.
50. The Best Friend of Charleston Railway Museum (www.bestfriendofcharleston.org).
51. S. Siles, *The Life of George Stephenson, Railway Engineer*, cited in Matt Ridley, *The Rational Optimist* (New York: HarperPerennial, 2010), pp. 283–84.
52. J. G. Martin, *Seventy-Five Years' History of the Boston Stock Exchange* (Boston, 1871), cited in Thies, "Development of the American Railroad Network," p. 16.

53. Holbrook, *Story of American Railroads*, p. 95.
54. Frances Carencross, *The Death of Distance* (Boston: Harvard Business School Press, 1997), coined the phrase "the death of distance" in reference to long-distance telecommunications. It is used here as an expression of the impact of railroads on the historically prevailing force of geography.
55. Taylor, *Transportation Revolution*, chart, p. 79: 3,328 miles of railroads and 3,326 miles of canals.
56. Stover, *Routledge Historical Atlas*, p. 20.
57. Gordon, *Empire of Wealth*, p. 151.
58. Rudi Volti, *Society and Technological Change* (New York: St. Martin's Press, 1955), p. 17.
59. Martin, *Railroads Triumphant*, p. 219.
60. Faith, *World the Railways Made*, pp. 129–30.
61. Ibid., p. 134.
62. Walter A. McDougall, *Throes of Democracy* (New York: HarperCollins, 2008), p. 148.
63. Ibid., p. 148.
64. Jonathan Hughes and Louis P. Cain, *American Economic History,* 7th ed. (New York: Pearson, 2007), p. 160–61.
65. Gordon, *Passage to Union*, p. 272.
66. Opinion of Hon. John M. Read, Supreme Court of Pennsylvania, *In Favor of the Passenger Cars Running Every Day of the Week, Including Sunday* (Philadelphia, 1867), cited in Gordon, *Passage to Union,* p. 114.
67. Stover, *Routledge Historical Atlas*, p. 44.
68. Holbrook, *Story of American Railroads*, p. 357.
69. Samuel Smiles, *The Life of George Stephenson, Railway Engineer* (Follett, Foster, and Co. 1859), p. 205. Interestingly, just as "paperless" personal computers led to a rise in the consumption of paper, the iron horse prompted an increase in its animal equivalent as wagons, coaches, and the like were needed to transport the rail delivery to its final destination. McDougall, *Throes of Democracy*, p. 150.
70. Seymour Dunbar, *A History of Travel in America*, vol. 3 (Indianapolis: Bobbs-Merrill Co., 1915), p. 938, quoting the Vincennes *Western Sun,* July 24, 1830.
71. Wolmar, *Blood, Iron, and Gold*, p. 78.
72. Michael Freeman, *Railways and the Victorian Imagination* (Yale University Press, 1999), p. 16.
73. Martin, *Railroads Triumphant*, p. 49.
74. Wolmar, *Blood, Iron, and Gold*, p. 91.

75. James W. Ely Jr., "Lincoln and the Rock Island Bridge Case" (Indianapolis: Indiana Historical Society).
76. Ibid., p. 8.
77. *Hurd v. Rock Island Railroad Company,* U.S. Circuit Court, Northern District of Illinois, August 1857.
78. Ely, "Lincoln and the Rock Island Bridge Case," p. 9.
79. Douglas, *All Aboard!*, p. 96.
80. "Our peculiar institution" was a southern euphemism for slavery.
81. Gordon, *Passage to Union*, p. 134.
82. Stover, *Routledge Historical Atlas*, pp. 46–47.
83. Albert J. Churella, *The Pennsylvania Railroad*, vol. 1 (University of Pennsylvania Press, 2012), p. viii.
84. Gordon, *Empire of Wealth*, p. 148.
85. Tony Judt, "The Glory of the Rails," *New York Review of Books*, December 23, 2010, and January 13, 2011.
86. Doron Swade, *The Difference Engine: Charles Babbage and the Quest to Build the First Computer* (New York: Viking, 2000), p. 10.
87. Anthony Hyman, *Charles Babbage, Pioneer of the Computer* (Princeton University Press, 1982), p. 143.
88. Swade, *Difference Engine,* p. 10.
89. Ibid., pp. 28–30.
90. Hyman, *Charles Babbage*, p. 165, quoting Babbage's *Passages from the Life of a Philosopher* (1864).
91. Ibid., p. 164.
92. Swade, *Difference Engine,* pp. 114–15.
93. Ibid., p. 306.
94. John Markoff, "It Started Digital Wheels Turning," *New York Times,* November 7, 2011.

Chapter 4

1. "A Monument to Charles Minot" (www.telegraph-history.org/charles-minot/index.html).
2. Ibid.
3. Albro Martin, *Railroads Triumphant: The Growth, Rejection & Rebirth of a Vital American Force* (Oxford University Press, 1992), pp. 23, 24.
4. Kenneth Silverman, *Lightning Man: The Accursed Life of Samuel F. B. Morse* (New York: Alfred A. Knopf, 2003), p. 73; Tom Standage, *The Victorian*

Internet: The Remarkable Story of the Telegraph and the Nineteenth Century's On-line Pioneers (New York: Berkeley Books, 1999), p. 25.

5. Carleton Mabee, *The American Leonardo: A Life of Samuel F. B. Morse*, rev. ed. (Fleischmanns, N.Y.: Purple Mountain Press, 2000), p. 98. Morse's nickname for his much younger (by eight years) bride was Lucrece.
6. Silverman, *Lightning Man*, pp. 73–74.
7. Ibid., p. 154.
8. Mabee, *American Leonardo*, p. 151.
9. Silverman, *Lightning Man*, p. 148.
10. Silverman, *Lightning Man*, pp. 148–49.
11. Mabee, *American Leonardo*, p. 154.
12. Other developments were also essential to the telegraph. Luigi Galvani's 1780 electric cell became Alessandro Volta's 1800 battery capable of maintaining a constant current. Dane Hans Christian Ørsted's 1820 discovery of the link between electricity and magnetism, and André-Marie Ampère's implementation of that link in Paris the same year, laid the scientific foundation exploited by the developers of the telegraph.
13. John Desmond Bernal, *A History of Classical Physics* (New York: Barnes and Noble Books, 1997), p. 284.
14. Standage, *Victorian Internet,* p. 17.
15. Mabee, *American Leonardo*, p. 192.
16. "Looking for the Electric Telegraph" (www.connected-earth.com); Paul DeMarinis, "The Messenger," 1998 (www.well.com/user/demarini/messenger.html).
17. Richard R. John, *Network Nation: Inventing American Telecommunication* (Belknap Press of Harvard University Press, 2010), p. 46.
18. Mabee, *American Leonardo*, p. 191.
19. Standage, *Victorian Internet,* p. 9. The message was "Si vous réussissez, vous serez bien-tôt couvert de gloire" (If you succeed, you will bask in glory).
20. Standage, *Victorian Internet,* p. 13.
21. John, *Network Nation*, p. 54.
22. Each wire was connected to a joystick-like controller. When the stick was vertical, the circuit was open and the needle was vertical. Move the stick to the left and the arrow/needle followed; move to the right and the current's direction reversed and the needle went right.
23. J. L. Kieve, *Electric Telegraph: A Social and Economic History* (Newton Abbott, U.K.: David & Charles, 1973), pp. 18–26.
24. Richard R. John, "The Selling of Samuel Morse," *Invention & Technol-*

ogy, Spring 2010. Interestingly, Morse held off his response to the February request for information until September 27: Mabee, *American Leonardo*, p. 196.

25. Mabee, *American Leonardo*, p. 183.
26. Silverman, *Lightning Man*, p. 160.
27. Ibid., p. 161.
28. Ibid., pp. 159–60; Mabee, *American Leonardo*, p. 192.
29. Silverman, *Lightning Man*, p. 153.
30. Mabee, *American Leonardo*, p. 311.
31. Silverman, *Lightning Man*, p. 168.
32. Mabee, *American Leonardo*, p. 207.
33. Ibid.
34. Silverman, *Lightning Man*, p. 169.
35. Ibid., pp. 196–98.
36. Ibid., p. 212. The British Wheatstone-Cooke team had been granted a U.S. patent for their needle telegraph eight days earlier.
37. Mabee, *American Leonardo*, p. 251. Fortunately for Morse, Cohen had agreed to a success-based fee.
38. Seymour Dunbar, *A History of Travel in America*, vol. 3 (Bobbs-Merrill, 1915), p. 1048.
39. Mabee, *American Leonardo*, p. 259.
40. *Journal of the House of Representatives,* December 30, 1840, p. 118.
41. Silverman, *Lightning Man*, pp. 219–21.
42. "Newspaper Accounts Regarding the Telegraph," *Industrial Revolution Reference Library,* vol. 3, *Primary Sources*, ed. James L. Outman, Matthew May, and Elisabeth M. Outman (Detroit: U-X-L Thomson Gale, 2003), p. 83.
43. *Congressional Globe*, 27th Congress, 3rd Session, p. 323.
44. Silverman, *Lightning Man*, p. 221. Representative Lew Wallace, Civil War hero of the Battle of Monocacy and author of *Ben-Hur,* attributed his subsequent electoral loss to his constituents' disdain for his spending their money on such a ridiculous idea. Lew Wallace, *An Autobiography,* 1906, p. 6, cited in Mabee, *American Leonardo*, p. 258.
45. *Congressional Globe*, 27th Congress, 3rd Session, p. 387.
46. "Inflation Calculator," DaveManuel.com (http://www.davemanuel.com/inflation-calculator.php).
47. Although the agreement with the railroad provided the B&O could use the telegraph free of charge, that capability went unrealized until after Charles Minot's experience.
48. Silverman, *Lightning Man*, p. 225.

49. Mabee, *American Leonardo*, p. 267.
50. Ultimately, Smith pocketed his half while Morse credited his to the government as a cost saving.
51. Robert Luther Thompson, *Wiring a Continent: The History of the Telegraph Industry in the United States, 1832–1866* (Princeton University Press, 1947), p. 22.
52. Silverman, *Lightning Man*, p. 230.
53. Lewis Coe, *The Telegraph: A History of Morse's Invention and Its Predecessors in the United States* (Jefferson, N.C.: McFarland & Co., 1993), p. 23.
54. Silverman, *Lightning Man*, pp. 230–31.
55. "The Selling of Samuel Morse," *Invention & Technology*, Spring 2010, pp. 45–46.
56. In a May 24 letter to his brother Sidney, Morse enclosed a news item he wanted placed in the *Journal of Commerce* in New York. In that description, Morse closed the quotation with a question mark.
57. Morse presented Anne Ellsworth with a copy of the historic tape containing the message she had selected. On this version, he hand-wrote a question mark at the end. Knowing of his affection for Anne, was it some kind of lovers' secret message inquiring as to their future together? No one will ever know. See John, *Network Nation*, p. 52.
58. Wright's preference was to run for governor of New York, which he did successfully in 1844, serving until 1846.
59. Mabee, *American Leonardo*, p. 279.
60. Report of the Postmaster General, Ex. Doc. No. 2, 29th Congress, 1st Session, p. 860.
61. Ibid., p. 861.
62. *Congressional Globe*, 28th Congress, 2nd Session, 1344–45, p. 366.
63. Jill Hills, *The Struggle for Control of Global Communication: The Formative Century* (University of Illinois Press, 2002), p. 29.
64. Thompson, *Wiring a Continent*, 240–41, chart. Although precise figures aren't available, the 1852 Census Report states that an additional 10,000 miles were under construction.
65. Henry David Thoreau, *Walden* (1854; New York: Dover Publications, 1995), p. 34.
66. German author Heinrich Heine, cited in Nicholas Faith, *World the Railways Made*, p. 42.
67. Walter A. McDougall, *Throes of Democracy* (New York: HarperCollins, 2008), p. 106.
68. Thompson, *Wiring a Continent*, p. 29.

69. Standage, *Victorian Internet,* p. 52.
70. *National Police Gazette,* May 30, 1846, cited in Silverman, *Lightning Man*, p. 240.
71. Jeffrey Sconce, *Haunted Media: Electronic Presence from Telegraphy to Television* (Duke University Press, 2000), p. 12.
72. Irwin Lebow, *Information Highways & Byways* (New York: IEEE Press, 1995), p. xiii.
73. Menahem Blondheim, *News over the Wires: The Telegraph and the Flow of Public Information in America, 1844–1897* (Harvard University Press, 1944), p. 33.
74. Standage, *Victorian Internet*, p. 149.
75. Richard DuBoff, "The Telegraph in Nineteenth-Century America: Technology and Monopoly," *Journal of the Society for Comparative Study of Society and History* 26, no. 4, (October 1984), p. 574.
76. Paul Starr, *The Creation of the Media: Political Origins of Modern Communications* (New York: Basic Books, 2004), p. 172.
77. JoAnn Yates, "The Telegraph's Effect on Nineteenth Century Markets and Firms," *Business and Economic History,* 2nd series, vol. 15 (1986), pp. 149–50.
78. For a further discussion, see Alfred Chandler, *The Visible Hand* (Belknap Press of Harvard University Press, 1977).
79. For a full discussion of Lincoln and the telegraph, see Tom Wheeler, *Mr. Lincoln's T-Mails: The Untold Story of How Abraham Lincoln Used the Telegraph to Win the Civil War* (New York: HarperCollins, 2006).
80. *Report of the Commissioner of Patents for the Year 1849*, pp. 489–90 (https://archive.org/stream/reportofcommissiunit#page/n495).
81. T. A. Watson, *Exploring Life* (Appleton Books, 1926), p. 68, cited in Brian Winston, *Media Technology and Society: A History from the Telegraph to the Internet* (New York: Routledge, 1998), p. 45.
82. Watson, *Exploring Life*, p. 71, cited in Winston, *Media Technology and Society*, p. 46.
83. Lebow, *Information Highways & Byways*, p. 35.
84. Seth Shulman, *The Telephone Gambit: Chasing Alexander Graham Bell's Secret* (New York: W. W. Norton, 2008), p. 12.
85. Ibid., pp. 12–13.
86. The Bell patent was actually filed on February 14, 1876, several weeks before the "Mr. Watson" breakthrough Bell recorded in his notebook. An intriguing and thoughtful discussion of whether Bell actually patented an invention he had not made in order to nose out others he knew were pursuing the same challenge is in Shulman, *The Telephone Gambit.*

Part III

1. Jonathan Hughes and Louis P. Cain, *American Economic History,* 7th ed. (Glenview, Ill.: Pearson, 2007), pp. 287, 356.
2. *Wikipedia*, s.v. "Western Union."
3. Paul Starr, *The Creation of the Media: Political Origins of Modern Communications* (New York: Basic Books, 2004), p. 202.

Chapter 5

1. Clark R. Mollenhoff, *Atanasoff: Forgotten Father of the Computer* (Iowa State University Press, 1988), p. 157.
2. Ibid.
3. Ibid., p. 158.
4. Jane Smiley, *The Man Who Invented the Computer* (New York: Doubleday, 2010), p. 64.
5. For a comprehensive study of the Atanasoff story, see Alice Rowe Burks, *Who Invented the Computer? The Legal Battle That Changed Computing History* (Amherst, N.Y.: Prometheus Press, 2003).
6. Doron Swade, *The Difference Engine* (New York: Viking, 2000), p. 10.
7. About twenty years prior to Pascal (circa 1623), the German Wilhelm Schickard had apparently constructed a "calculating clock" using rods and cylinders to render a conclusion to input data. Unfortunately, Schickard became a victim of the plague; his ideas were discovered only recently through his correspondence with another contemporary mathematician.
8. Since the gears rotated in only one direction, subtraction was accomplished by a mathematical trick called the "nines complements," the difference between any number and a similar collection of nines. For instance, to subtract 500 from 800, the Pascaline had a lever that switched the dialed number to its nines complement. Dial in 500 and the complement of 499 (999 - 500) would be displayed, then returning the lever to its regular position, the user dialed in 800 to add it to 499, producing 1299; then the user took the furthest left number (in this case 1) and added it to the others. Voila! The answer to 800 - 500 is 300!
9. The "carry" function was the Pascaline's bête noire because it would frequently jam. Thirty years later the German Gottfried Wilhelm von Leibniz substituted a cylinder with nine rows of teeth to pass the carry to the next column. It was this model Babbage followed. Leibniz named his device the "step reckoner." Because it dispensed with Pascal's carry lever, the step reckoner's gears could

also move in both directions, making subtraction possible without the awkward use of the nines complement. See Stan Augarten, *Bit by Bit: An Illustrated History of Computers* (New York: Ticknor & Fields, 1984).

10. As so often in the development of technology, Babbage was not the only person with such an insight, In the late 1700s a German army captain, J. H. Müller, had a similar idea and petitioned his government for funds to build it. When the funding was not forthcoming, the project died.
11. Funding problems and a dispute with the man he hired to build the engine resulted in only a 24-inch-high, 19-inch-wide, and 14-inch-deep portion of the engine being built. As we have seen in chapter 3, Babbage's difference engine was ultimately built in 1991 by the London Science Museum. It performed as Babbage forecast.
12. James Gleick, *The Information: A History, a Theory, a Flood* (New York: Pantheon Books, 2011), p. 101.
13. Ibid., p. 114.
14. His son, Henry, assembled a part of it (a portion of the mill) in 1889.
15. One person who did not find Babbage's ideas quirky was Ada Lovelace, daughter of the English poet Lord Byron. An emancipated woman in a male-dominated era, Ada had a love for mathematics that brought her into contact with Babbage. Their correspondence ultimately produced its own history-making result. In her footnotes to a paper about Babbage's work that she translated from Italian, she added commentary on the manner in which information could be entered into the analytical engine. While there is dispute as to whether Babbage contributed to these observations, it remains the first written description of what today we call "software." Lovelace also recognized that the machine had practical implications beyond mathematical calculations. "The engine can arrange and combine its numerical quantities exactly as if they were letters or any other general symbols," she wrote. Steven Johnson, *How We Got to Now* (New York: Riverhead Books, 2014), p. 248.
16. Augarten, *Bit by Bit*, p. 82. By 1913 Burroughs employed 2,500 people.
17. Eric G. Swedin and David L. Ferro, *Computers: The Life Story of a Technology* (Johns Hopkins University Press, 2005), pp. 20–21.
18. James Burke, *Connections* (New York: Simon & Schuster Paperbacks, 1995), p, 111.
19. Augarten, *Bit by Bit*, p. 145.
20. George B. Dyson, *Darwin among the Machines: The Invention of Global Intelligence* (New York: Basic Books, 1997), pp. 53–58. When World War II broke out, Turing's mathematical skills brought him to the British codebreaking operation at Bletchley Park. While electromechanical devices were

used early in the war to help break the German codes, they weren't really Turing machines. By 1943, technology had advanced to the much-heralded Colossus, "totally electronic binary analytical calculators that were structurally designed to solve logical problems, to give plain-English read outs of encrypted texts, and to reconstruct the processes and key or keys of the German cipher system." Georges Ifrah, *The Universal History of Computing* (New York: John Wiley & Sons, 2001), p. 218.

21. Augarten, *Bit by Bit*, p. 89.
22. Because it used telephone relays it was an electromechanical device, but Zuse and his colleague Helmut Schreyer had envisioned an all-electronic vacuum-tube-based machine.
23. Ifrah, *Universal History of Computing*, p. 206.
24. Interestingly, this did not include Atanasoff's use of binary math and Boolean logic, which would have simplified the task.
25. Amazingly, Mauchly would visit Atanasoff at his Ordnance Lab post during the development of ENIAC to continue picking his brain.
26. Swedin and Ferro, *Computers*, p. 39.
27. In this effort Mauchly and Eckert were aided greatly by John von Neumann, a Hungarian émigré, Princeton professor, and part of the Manhattan Project.
28. Even though work on the program started in 1945, the EDVAC wasn't completed until 1952, after the UNIVAC.
29. The company was offered to IBM but the firm's lawyers nixed the deal out of fear it wouldn't pass antitrust scrutiny since IBM controlled a dominant position in mechanical tabulators (the legacy of Hollerith) and calculators.
30. *Sperry Rand Corporation et al. v. Bell Telephone Laboratories, Inc.*, 317 F. 2d 491 (2d Cir. 1963).
31. *Honeywell, Inc. v. Sperry Rand Corp, et al.*, 180 USPQ 673 (D. Minn 1973).
32. Burks, *Who Invented the Computer?*, p. 13.
33. Shockley shared the prize with John Bardeen and Walter Brattain, also of Bell Labs.
34. Wade Rowland, *Spirit of the Web* (Toronto: Somerville House Publishing, 1997), p. 311.
35. "Flashback: The History of Computing," *Computerworld,* May 19, 1999.
36. Jack Kilby of Texas Instruments received the 2000 Nobel Prize in Physics for the 1959 invention of the integrated circuit.
37. Noyce's development, for instance, relied on the discovery of another of the "Fairchild Eight," Jean Hoerni, who conceived of the means of making a completely flat transistor (earlier versions had a tiny mesa protruding)—the planar process.

38. Intel Corporation, "The History of Intel, 30 Years of Innovation," Intel.com, 2002.
39. Ibid.
40. Ibid.
41. Ron Smith, Intel senior vice president, to author, March 2002.
42. Rowland, *Spirit of the Web*, p. 322.
43. For a user-friendly discussion of microprocessors, see www.intel.com/education/teachtech/learning/chips/index.htm.
44. While the growth of microprocessor sales was slow, Intel's broadened scope of activity turned out to be a fortunate decision. In the late 1970s the bottom fell out of DRAM profitability because of inexpensive foreign production. Intel was losing its shirt on DRAM and left the business in 1985. By that time, however, the microprocessor was the backbone of the IBM PC, and Intel was off to the races.
45. The Frenchman Andre Thi Truong had created a PC, the Micral, in 1973 and sold 500 units. Its design was never published in the United States. For a discussion of the development of the PC, see Swedin and Ferro, *Computers*, chap. 5.
46. Swedin and Ferro, *Computers*, pp. 88–89.
47. Ibid., p. 93.
48. Augarten, *Bit by Bit*, p. 280.
49. The company was NABU: The Home Computer Network, the first network to connect PCs to cable TV lines.
50. Gordon Moore, "Cramming More Components onto Integrated Circuits," *Electronics,* April 19, 1965.

Chapter 6

1. Bell Labs was then located at 463 West Street in Manhattan. It would not move to New Jersey until after World War II.
2. There are two parts to a telephone call: the transmission of the signal and the setup of the circuit that carries it. The relay automated the setup activity formerly performed by switchboard operators pulling cables to plug two lines together. In the "on" position, the relay acted like the operator plugging the lines together; the "off" position broke the circuit path. Unknown to him, Stibitz was demonstrating what Konrad Zuse was building in Germany at the same time.
3. Paul E. Ceruzzi, *Reckoners: The Prehistory of the Digital Computer, from*

Relays to Stored Program Concept, 1935–1945 (Westport, Conn.: Greenwood Press, 1983).

4. The phone lines were actually twenty-eight-wire teletype cable.
5. It would be ten years before remote computing would happen again at the National Bureau of Standards (today the National Institute for Standards and Technology), with the Standards Eastern Automatic Computer (SEAC).
6. Richard R. John, *Network Nation* (Belknap Press of Harvard University Press, 2010), p. 159, citing Orton correspondence in E. C. Baker, *Sir William Preece, F.R.S., Victorian Engineer Extraordinary* (London: Hutchinson, 1976).
7. By threatening to use Edison's patent, Gould depressed Western Union stock with the threat that multiple simultaneous signals would drive down rates and thus threaten Western Union's revenues. Thomas Edison used the proceeds from the sale to Gould to build his legendary Menlo Park laboratories.
8. Hubbard had made his mark dealing in horse-drawn trolleys, water and gas works, and other activities that required a municipal franchise. He was also a champion of nationalizing the telegraph monopoly.
9. John, *Network Nation*, p. 164. Another father of a Bell student, Thomas Sanders, supported his research and the early stages of the telephone business as well.
10. Herbert Newton Casson, *The History of the Telephone* (Chicago: A. C. McClurg, 1910), p. 8.
11. There has been much debate over the patent caveat (a declaration of an idea with the patent application to follow) filed the same day as Bell's application by Elisha Gray of Chicago. It is an interesting story of competing concepts that is well told in Seth Shulman's *The Telephone Gambit* (New York: W. W. Norton, 2008). After approximately 600 lawsuits challenging the patent, and a decision of the Supreme Court, the Bell patent prevailed.
12. The first record of the idea of transmitting sound over wire was an 1861 lecture to the Frankfurt Physical Society by the German inventor Philip Reis. Vaclav Smil, *Creating the Twentieth Century: Technical Innovations of 1867–1914 and Their Lasting Impact* (Oxford University Press, 2005), p. 228.
13. Casson, *History of the Telephone,* p. 13.
14. Ibid., p. 16.
15. John, *Network Nation*, p. 162.
16. Casson, *History of the Telephone,* p. 21.
17. Tim Wu, *The Master Switch: The Rise and Fall of Information Empires* (New York: Alfred A. Knopf, 2010), p. 29.

18. Hubbard had met Vail while serving on President Hayes's commission on mail transportation: Casson, *History of the Telephone*, p. 23.
19. James Gleick, *The Information: A History, a Theory, a Flood* (New York: Pantheon Books, 2011), p. 189, with citation.
20. Smil, *Creating the Twentieth Century*, p. 233.
21. Wu, *Master Switch*, p. 26.
22. Ibid., p. 27.
23. Hubbard had also hired lawyers to file suit attacking Western Union's patent claims.
24. Wu, *Master Switch*, p. 31.
25. Casson, *History of the Telephone*, p. 30.
26. Ibid., p. 30. The Bell patent wars continued until in 1888, in a 4-3 decision, the U.S. Supreme Court held that Alexander Graham Bell had indeed invented the telephone.
27. Ibid., p. 61.
28. Ibid., p. 62.
29. Wu, *Master Switch*, p. 46, citing Horace Coon, *American Tel & Tel: The Story of a Great Monopoly* (1939; Books for Libraries Press, 1971), pp. 66–67.
30. The corporate name was changed to AT&T in 1885 by then president William Forbes.
31. Coon, *American Tel & Tel*, p. 102
32. Ibid., p. 103.
33. John, *Network Nation*, p. 340.
34. Jon Gertner, *The Idea Factory: Bell Labs and the Great Age of American Innovation* (New York: Penguin, 2012), p. 19.
35. Ibid., p. 127.
36. Wu, *Master Switch*, pp. 104–06.
37. Tariff FCC No. 132, filed April 16, 1957.
38. The total control over the network exercised by the Bell System sometimes bordered on the humorous. The idea that "foreign attachments" could endanger the network seems far-fetched in this age when a trip to the drugstore brings back a telephone that easily attaches to the network with no deleterious effects. In the ultimate illustration of thwarting consumers in the name of "one policy, one system, one universal service," a plastic phonebook cover containing local advertising was deemed a "foreign attachment" and banned because it covered (and competed with) the advertising in the Bell-owned *Yellow Pages*, whose revenues, it was argued, helped to reduce the cost of phone service.
39. Even then, however, the federal government allowed AT&T to design com-

plex and expensive tests that had to be passed before a modem was allowed to connect. One result of this was to force the introduction of acoustically coupled (as opposed to electronically coupled) modems, those silly-looking boxes with rubber ears into which fit a telephone receiver.

40. Paul Baran, "On Distributed Communications" (Santa Monica, Calif.: RAND Corporation, 1964).
41. Paul Baran explained, "In the analog days both ends of the connection needed to work in tandem, and the probability of many things working in tandem without failing was so low that you had to make every part nearly perfect. But if you don't care about reliability any more [because the packets simply reroute themselves around the fault], then the cost of the components goes way down." Quoted in Stewart Brand, "Founding Father," *Wired*, March 1, 2001.
42. Ibid.
43. Paul Baran, conversation and January 29, 2011, email with the author regarding Baran's attempts to explain packet-switching concepts to executives and engineers whose only experience had been with circuit-switched networks.
44. Katie Hafner and Matthew Lyon, *Where Wizards Stay Up Late: The Origins of the Internet* (New York: Touchstone, 1996), p. 64.
45. The first ARPANET nodes were at Stanford Research Institute, UCLA, UCSB, and the University of Utah.
46. In 1965 Donald Davies, the Englishman who coined the term "packet switching," proposed something similar to ARPANET at the National Physical Laboratory in the United Kingdom. It never was funded, but many of his ideas found their way into ARPANET.
47. The key to the early ARPANET was the installation at each site of a new computer, an interface message processor (IMP) that would control the network connection (send and receive data, check for errors, route the data, and so on). A common protocol was established for how the IMPs would communicate with their host computers. See Hafner and Lyon, *Where Wizards Stay Up Late*, p. 75.
48. Carolyn Duffy Marsan, "The Evolution of the Internet," *NetworkWorld,* February 9, 2009.
49. While ARPANET was only for noncommercial uses, the edges of that rule began to blur as companies such as Hewlett-Packard connected their network, albeit for the purpose of furthering the research HP was conducting. By the time ARPANET was decommissioned in 1990 the internet working of TCP/IP knew no commercial/noncommercial bounds.
50. Others involved included Bob Braden, Jon Poster, and Steve Crocker.

51. TCP/IP has an abstract construct that represents different activities as layers in the "network stack." At the bottom layer is the physical medium (for example, optical fiber) over which the message travels. One step up is the link layer, which describes how the message will be sent over the physical layer (for example, the protocol for Wi-Fi). Next up the stack is the internet layer, described above. At the top of the stack is the application layer containing the actual message.
52. Of course, there is also the incremental cost of building added capacity. That, too, is decreasing as a result of digitization. AT&T, for instance, reported that "in 2015/16 we're going to deploy about 250% of the capacity that we did in 2013/14, and we're going to do it for 75% of the cost." "AT&T's (T) Management Presents at Wells Fargo 2016 Convergence & Connectivity Symposium (Transcript)," SeekingAlpha.com, June 21, 2016.
53. Radio was originally envisioned as an application of the telephone network, but point-to-multipoint broadcasting proved much more efficient than a circuit occupying point-to-point wired connection. When television signals first began to be distributed by coaxial cable (a technology available to the telephone company), it was opposed by phone companies, many of which denied the cable access to their telephone poles.
54. Hal Varian, quoted in Steven Levy, *How Google Thinks, Works and Shapes Our Lives* (New York: Simon & Schuster, 2011), p. 117.

Chapter 7

1. International Telecommunications Union (ITU), "ITU Releases Annual Global ICT Data and ICT Development Index Country Rankings," November 30, 2015 (http://www.itu.int/net/pressoffice/press_releases/2015/57.aspx#.VqUXuTFdGj2).
2. Ibid.
3. Zachery Davies Boren, "There Are Officially More Mobile Devices Than People in the World," *The Independent,* October 7, 2014. This does not mean that every person has a mobile phone. However, it is a measure of the scope and scale of wireless penetration.
4. While texting and some rudimentary data services were offered on some devices in 2002, it was still principally a mobile *phone,* and other applications were ancillary.
5. Irwinb Lebow, *Information Highways & Byways* (New York: IEEE Press, 1995), p. 76.
6. Albert Bigelow Paine, *In One Man's Life: Being Chapters from the Personal &*

Business Career of Theodore N. Vail (1921; London: Forgotten Books, 2012), p. 275.

7. The transmission moved by wire to Arlington, Virginia, where it was flung into the air by AT&T's new transmission station.
8. Paine, *In One Man's Life*, p. 281.
9. Ibid., p. 283.
10. Jon Gertner, *The Idea Factory: Bell Labs and the Great Age of American Innovation* (New York: Penguin, 2012), p. 297.
11. Tom Farley, "The Cell-Phone Revolution," *American Heritage* 22, no. 3 (Winter 2007).
12. Gertner, *Idea Factory*, p. 280.
13. Ibid., p. 281.
14. Farley, "Cell-Phone Revolution."
15. As a result of the antitrust lawsuit *United States v. AT&T,* the company entered a consent decree with the U.S. Department of Justice on January 8, 1982, under which on January 1, 1984, the company would divest itself of the regional operating companies.
16. Farley, "Cell-Phone Revolution."
17. Robert Roche, vice president, research, CTIA: The Wireless Association; email correspondence with author, August 12, 2009.
18. S. J. Blumberg and J. V. Luke, "Wireless Substitution: Early Release of Estimates from the National Health Interview Survey, January–June 2012," National Center for Health Statistics, December 2012 (http://www.cdc.gov/nchs/nhis.htm).
19. Study by Robert Jensen of Harvard of coastal village of Kerala, *The Economist*, Special Report, September 26, 2009, p. 7.
20. "A Doctor in Your Pocket, Health Hotlines in Developing Countries," GSMA Development Fund, 2009.
21. "Airtime Is Money," *The Economist*, January 19, 2013.
22. Jokko Initiative: Mobile Technology Amplifying Social Change (www.tostan.org).
23. Geneva Health Forum (www.ghf10.org/reports/209).
24. World Bank study by Christine Zhan-Wei Qiang, *The Economist*, Special Report, September 26, 2009, p. 7.
25. A discussion of One Laptop Per Child can be found at http://laptop.org.
26. Cisco Corporation, "Cisco Visual Networking Index: Global Mobile Data Forecast Update, 2016–2021," Cisco.com, March 29, 2017 (www.cisco.com/c/en/us/solutions/collateral/service-provider/visual-networking-index-vni/mobile-white-paper-c11-520862.html).

27. Henry David Thoreau, *Walden* (1854; New York: Dover, 1995), p. 34.
28. "The Mobiles: Social Evolution in a Wireless Society," Context Research Group, 2002.
29. Rich Ling, *New Tech, New Ties: How Mobile Communication Is Reshaping Social Cohesion* (MIT Press, 2008).
30. Rich Ling, *The Mobile Connection: The Cell Phone's Impact on Society* (Burlington, Mass.: Morgan Kauffman, 2004), p. 214n.
31. Tolu Oguniesi, "Seven Ways Mobile Phones Have Changed Lives in Africa," CNN, September 13, 2012.
32. Theodore Caplow, Louis Hicks, and Benjamin J. Wattenberg, *The First Measured Century* (Washington, D.C. AEI Press, 2001), p. 24.
33. Judith Flanders, *The Making of Home* (New York: St. Martin's Press, 2016), p. 49.
34. The concept of "historical reintegration" was first proposed by James Katz of Rutgers University. See *The Economist,* April 12, 2008.
35. Paine, *In One Man's Life*, p. 283.
36. Walter A. McDougall, *Throes of Democracy* (New York: HarperCollins, 2008), p. 150 (cows and hens); Matt Ridley, *The Rational Optimist* (New York: HarperPerennial, 2010), p. 283.
37. Judith Nicholson, "Sick Cell: Representations of Cellular Telephone Use in North America," *Journal of Media and Culture* 4, no. 3 (June 2001).
38. Denise Grady, "Cellphones Are Still Safe for Humans, Researchers Say," *New York Times,* February 2, 2018.
39. Mark Weiser, "The Computer for the 21st Century," *Scientific American* 265, no. 3 (September 1991), pp. 66–75.
40. "eCall—Saving Lives through In-Vehicle Communication Technology," European Commission, August 2009.
41. Roxie Hammil and Mike Hendricks, "Gadgets to Help Tend a Garden," *New York Times*, April 24, 2013.
42. Don Clark, "Take Two Digital Pills and Call Me in the Morning," *Wall Street Journal,* August 4, 2009.
43. "New High-Tech Sensor-Laden Carpet May Revolutionize Building Security," *Defense Review*, April 15, 2005.

Chapter 8

1. Rhoda Thomas Tripp, *The International Thesaurus of Quotations* (New York: T. Y. Crowell Co., 1970), p. 280.

2. Jeremy Rifkin, *The Zero Marginal Cost Society* (New York: Palgrave Macmillan, 2014).
3. The topic of open networks is an essential predicate to optimizing these new networks. The most divisive issue during my tenure at the FCC was the Open Internet Order, which imposed on internet service providers the obligation to behave as a nondiscriminatory common carrier and carry all traffic free of blocking, throttling, or payment for priority handling. Absent an open internet, network operators become gatekeepers free to discriminate to advantage themselves. Unfortunately, the Trump FCC has eliminated those protections.
4. The typical telegraph operator could transmit at 3 bits per second; 25,000,000 bits/sec. ÷ 3 bits/sec. = 8.33 million-fold increase in speed.
5. Farhad Manjoo, "How Amazon's Long Game Yielded a Retail Juggernaut," *New York Times,* November 18, 2015.
6. While the network may have created information such as boxcar loadings or the number of news dispatches sent, it was an ancillary by-product.
7. M. G. Siegler, "Eric Schmidt: Every 2 Days We Create As Much Information As We Did Up to 2003," *TechCrunch,* August 4, 2010.
8. "Data Age 2025" (https://www.idc.com/prodserv/custom-solutions/RESOURCES/ATTACHMENTS/thought-leadership-cs.pdf).
9. Jon Gertner, "Behind GE's Vision for the Industrial Internet of Things," *Fast Company,* June/July 2014.
10. Siegler, "Eric Schmidt: Every 2 Days."
11. David B. Angus, "Give Up Your Data to Cure Disease," *New York Times,* February 6, 2016.
12. Bobbie Johnson, "Privacy No Longer a Social Norm, Says Facebook Founder," *The Guardian,* January 10, 2010.
13. Samuel Warren and Louis Brandeis, "The Right to Privacy," *Harvard Law Review* 4 (1890), pp. 193–220.
14. Joseph Turow, Michael Hennessy, and Nora Draper, *The Tradeoff Fallacy: How Marketers Are Misrepresenting American Consumers and Opening Them Up to Exploitation,* a report from the Annenberg School for Communications, University of Pennsylvania, June 2015.
15. Although the Supreme Court has held privacy to be the central reason for the Fourth Amendment protection against search and seizure, and has asserted a so-called "constitutional right to privacy" (*Griswold v. Connecticut,* 381 U.S. 479), neither the word "privacy" nor the concept itself is ever explicitly mentioned in the U.S. Constitution.

16. One internet service provider, AT&T, appropriately discloses how it collects a staggering amount of information on each subscriber through "a combination of information from wireless and WiFi locations, TV viewing, calling and texting records, website browsing and mobile application usage and other information we have about you and other customers" (http://www.att.com/gen/privacy-policy?pid=13692).
17. Unfortunately, in the very early days of the Trump administration Congress repealed these protections.
18. "Technology at Work," *Citi Global Perspectives & Solutions*, February 2015, p. 16.
19. Cecilia Kang, "No One in Washington Is Talking about the Problems with the Sharing Economy, Except This Lawmaker," *Washington Post,* June 26, 2015.
20. Burson-Marsteller, The Aspen Institute Future of Work Initiative, *Time*, "Forty-Five Million Americans Say They Have Worked in the On-Demand Economy," January 6, 2016, polling by Penn Schoen Berland.
21. Thomas W. Malone and Robert Laubacher, "The Dawn of the E-Lance Economy," *Harvard Business Review,* September-October 1998.
22. Dave Masters, "Say Goodbye to the '9 to 5': Futurologist Makes Five Predictions for the Workplace of the Future," *Daily Record*, December 23, 2016.
23. See Antoine van Agtmael and Fred Bakker, *The Smartest Places on Earth: Why Rustbelts Are the Emerging Hotspots of Global Innovation* (New York: PublicAffairs, 2016).
24. Quentin Hardy, "Gearing Up for the Cloud, AT&T Tells Its Workers: Adapt or Else," *New York Times*, February 13, 2016.
25. J. H. A. Bone, "Old English Guilds and Trade Unions," *Atlantic Monthly* 39, no. 231 (March 1877), p. 284.
26. The Guild of Palmers of Ludlow (England) provided that "if any member becomes a leper, or blind, or maimed in limb, or smitten with any other incurable disorder (which God forbid!) we wish that the goods of the guild be largely bestowed on him." Bone, "Old English Guilds and Trade Unions," p. 284.
27. I first heard this great description expressed by Ong Ye Kung, Singapore's minister of education.
28. Flipping the teaching process has even reached the gridiron. Urban Meyer, head football coach at Ohio State University, has abandoned diagramming plays on the chalkboard. Now coaches send each player a video or graphical representation of what they are to learn and the players watch on their computers or mobile devices. Team meetings are used to ensure the players

understand the use of that information. Jonathan Clegg, "How Urban Meyer Took the Buckeyes to School," *Wall Street Journal,* December 7, 2014.

29. *Marketplace* staff, "Conversations about Mobility, Live from Aspen," *NPR Marketplace,* June 29, 2015, audio interview with Jose Ferreira, CEO of Knewton.
30. Alec Ross, senior adviser to Secretary of State Hillary Clinton, discussion with author.
31. Joshua Cooper Ramo, *The Seventh Sense* (New York: Little, Brown, 2016), p. 34.
32. "Estonia Is Putting Its Country in the Cloud and Offering Virtual Residency," *The Conversation,* March 26, 2017.
33. James Madison, *The Federalist No. 10* (Wesleyan University Press, 1961), pp. 57, 58.
34. The talk in which Ghonim makes these points is discussed by Thomas Friedman, "Social Media: Destroyer or Creator?," *New York Times*, February 3, 2016.
35. Porter Bibb, *It Ain't As Easy As It Looks* (New York: Crown Publishers, 1993), p. 180.
36. Mark R. Robertson, "300+ Hours of Video Uploaded to YouTube Every Minute," *TubularInsights: Video Marketing Insights* (YouTube video), November 21, 2014.
37. *The Daily Show with Jon Stewart,* "End Times," June 10, 2009.
38. Michael Barthel, Elisa Shearer, Jeffrey Gottfried, and Amy Mitchell, "The Evolving Role of News on Twitter and Facebook," Pew Research Center, July 14, 2015.
39. Eric Burns, *Infamous Scribblers* (New York: PublicAffairs, 2006), p. 3.
40. Author's conversation with Ron Nessen, November 28, 2015.
41. Tom Standage, *Writing on the Wall* (New York: Bloomsbury, 2013), p. 3.
42. James B. Stewart, "Facebook Has 50 Minutes of Your Time Each Day. It Wants More," *New York Times*, May 5, 2016.
43. Jonah Berger and Katherine L. Milkman, "What Makes Online Content Viral?," *Journal of Marketing Research* 49, no. 2 (April 2012), pp. 192–205.
44. Craig Silverman, "Viral Fake Election News Outperformed Real News on Facebook," *BuzzFeed*, November 16, 2016.
45. William J. Bernstein, *Masters of the Word: How Media Shaped History* (New York: Grove Press, 2013), p. 155.
46. Extreme poverty is defined as living on less than $1.90 per day. The number living in extreme poverty decreased from 1.75 billion people in 1990 to 702.1 million in 2015, or from 37.1 percent of the world population to 9.6 percent.

World Bank, *Global Monitoring Report 2015/2016* (www.worldbank.org/en /publication/global-monitoring-report).

47. "ITU Releases Annual Global ICT Data and ICT Development Index," press release, International Telecommunications Union, November 30, 2015.
48. Lawrence Yanovitch, executive director, GSMA Mobile for Development Foundation, conversation with the author, February 2016.
49. World Bank, *World Development Report 2016: Digital Dividends* (Washington, D.C.: World Bank, 2016).
50. International Telecommunications Union, "The Digital Divide in 2015," *ICT Facts and Figures,* May, 2015.
51. Adie Tomer and Joseph Kane, "Broadband Adoption Rates and Gaps in U.S. Metropolitan Areas," Brookings Institution, December 7, 2015.
52. "Homework gap" was a term coined by my colleague Commissioner Jessica Rosenworcel.
53. "Living in poverty" is defined as meeting eligibility requirements for free or reduced-price school lunches. See Lyndsey Layton, "Majority of U.S. Public School Students Are in Poverty," *Washington Post,* January 16, 2015.
54. Anton Troianovski, "The Web-Deprived Study at McDonald's," *Wall Street Journal,* January 28, 2013.
55. Kudos to Commissioner Mignon Clyburn for her tireless leadership and advocacy regarding the FCC's Lifeline program for low-income Americans. Unfortunately, when the Trump FCC came into office, it immediately began scaling back this program.
56. One proposal was for the ViaSat 3, with a throughput of one terabit per second.
57. OneWeb proposed 700 satellites and SpaceX 4,000—all made possible by the reduced cost of construction and the huge savings in putting them into orbit, thanks to the commercial orbital lift industry. Whereas earlier-generation satellites could cost $200 million for a custom-crafted vehicle, current-generation satellites cost less than $1 million and roll off an assembly line like automobiles.

Chapter 9

1. Robert Gordon, *The Rise and Fall of American Growth* (Princeton University Press, 2016), p. 14.
2. Ibid.
3. Ibid., p. 7.

4. kdespagniqz, "Connected Cars Will Send 25 Gigabytes of Data to the Cloud Every Hour," *Quartz,* February 13, 2015.
5. Patrick Nelson, "Just One Autonomous Car Will Use 4,000 GB of Data/Day," *Network World*, December 7, 2016.
6. Credit for the push-to-pull concept belongs to my friend Bill Coleman, former CEO of Veritas.
7. U.S. Environmental Protection Agency, "Water Audits and Water Loss Control for Public Water Systems," July 2013.
8. John McCarthy, "Ascribing Mental Qualities to Machines," Stanford University, 1979.
9. Ray Kurzweil, *The Singularity Is Near* (New York: Viking, 2006).
10. Field-programmable gate array (FPGA) microchips are the next processing iteration. While GPUs have their programming specifics built in, FPGAs, as the name suggests, are configured by a customer after manufacturing, thus increasing their flexibility and their potential applications.
11. Carl Benedikt Frey and Michael A. Osborne, "The Future of Employment: How Susceptible Are Jobs to Computerisation?," January 2013. 10.1016/j.techfore.2016.08.019 (https://www.researchgate.net/publication/271523899_The_Future_of_Employment_How_Susceptible_Are_Jobs_to_Computerisation).
12. Hearings before the Subcommittee on Employment and Manpower, U.S. Senate, part I, May 21, 22, 23, 1963, p. 321 (https://catalog.hathitrust.org/Record/100662611).
13. Tom Standage, "The Return of the Machinery Question," *The Economist,* Special Report, June 25, 2016.
14. Ibid., p. 9.
15. Craig Giffi, "The Skills Gap in US Manufacturing: 2015 and Beyond," Deloitte and the Manufacturing Institute, 2011.
16. Ruchir Sharma, "Robots Won't Kill the Workforce. They'll Save the Global Economy," *Washington Post,* December 2, 2016.
17. Vivek Wadhwa, "Why China Won't Own Next-Generation Manufacturing," *Washington Post,* August 26, 2016.
18. The "new occupations, not new jobs" theme was developed by my friend Susan Crawford of Harvard University. Conversation with the author, February 2017.
19. Giffi, "Skills Gap in US Manufacturing."
20. Quentin Hardy, "Gearing Up for the Cloud, AT&T Tells Its Workers: Adapt or Else," *New York Times,* February 13, 2016.

21. See McKinsey & Co., "How Blockchains Could Change the World," May 2016.
22. Igal Zeifman, "2015 Bot Traffic Report: Humans Take Back the Web, Bots Not Giving Any Ground," Incapsula.com, December 9, 2015.
23. Elias Groll, "Did Russia Knock Out Ukraine's Power Grid?," *Foreign Policy*, January 8, 2016.
24. Joseph Marks, "Indictment: Iranians Made 'Coordinated' Cyberattacks on U.S. Banks, Dam," *Politico,* March 24, 2016.
25. U.S. Department of Justice, Office of Public Affairs, "Rusian Cyber-Criminal Convicted of 38 Counts Related to Hacking Businesses and Stealing More Than Two Million Credit Card Numbers," August 25, 2016.
26. Josh Rogin, "NSA Chief: Cybercrime Constitutes the 'Greatest Transfer of Wealth in History,'" *Foreign Policy*, July 19, 2012.
27. "Former Cardinal Exec Sentenced to Jail for Hacking Astros," *Sports Illustrated*, July 18, 2016.
28. Anthony Spadafora, "The Average IoT Device Is Compromised after Being Online for 6 Minutes," ITPortal.com, October 18, 2016.
29. "Morning Cybersecurity," *Politico,* November 11, 2016.
30. To move beyond cyber catchup, I fought for a stipulation in the spectrum grants for fifth-generation wireless (5G) to require that cybersecurity be designed into the technology from the outset. The industry and its allies opposed it as overly burdensome. Nevertheless, it became the law. Unfortunately, it was subsequently repealed by the Trump FCC in a sad indication of the commitment level of both industry and government to address cybersecurity.
31. David Ignatius, "The Cold War Is Over: The Cyber War Has Begun," *Washington Post,* September 15, 2016.

Index